McDougal Littell
Pre-Algebra

Larson   Boswell   Kanold   Stiff

CHAPTER 9

Resource Book

The Resource Book contains a wide variety of blackline masters available for Chapter 9. The blacklines are organized by lesson. Included are support materials for the teacher as well as practice, activities, applications, and project resources.

McDougal Littell
A DIVISION OF HOUGHTON MIFFLIN COMPANY

Evanston, Illinois • Boston • Dallas

Contributing Authors

The authors wish to thank **Jessica Pflueger** for her contributions to the Chapter 9 Resource Book.

Copyright © 2005 by McDougal Littell, a division of Houghton Mifflin Company.
All rights reserved.

Permission is hereby granted to teachers to reprint or photocopy in classroom quantities the pages or sheets in this work that carry a McDougal Littell/Houghton Mifflin copyright notice. These pages are designed to be reproduced by teachers for use in their classes with accompanying McDougal Littell material, provided each copy made shows the copyright notice. Such copies may not be sold and further distribution is expressly prohibited. Except as authorized above, prior written permission must be obtained from McDougal Littell, a division of Houghton Mifflin Company, to reproduce or transmit this work or portions thereof in any other form or by any other electronic or mechanical means, including any information storage or retrieval system, unless expressly permitted by federal copyright laws. Address inquiries to Supervisor, Rights and Permissions, McDougal Littell, P. O. Box 1667, Evanston, IL 60204.

ISBN: 0-618-26946-0

3456789-QDI-09 08 07 06 05 04

Contents

CHAPTER 9 Real Numbers and Right Triangles

Chapter Support	1–4
9.1 Square Roots	5–13
9.2 Simplifying Square Roots	14–21
9.3 The Pythagorean Theorem	22–29
9.4 Real Numbers	30–37
9.5 The Distance and Midpoint Formulas	38–47
9.6 Special Right Triangles	48–55
9.7 The Tangent Ratio	56–64
9.8 The Sine and Cosine Ratios	65–72
Review and Projects	73–81
Resource Book Answers	A1–A7

Contents

CHAPTER SUPPORT MATERIALS

Tips for New Teachers	p. 1
Parents as Partners	p. 3

LESSON MATERIALS

	9.1	9.2	9.3	9.4	9.5	9.6	9.7	9.8
Lesson Plans (Reg. & Block)	p. 5	p. 14	p. 22	p. 30	p. 38	p. 48	p. 56	p. 65
Activity Master					p. 40			
Tech. Activities & Keystrokes	p. 7							
Practice A	p. 8	p. 16	p. 24	p. 32	p. 41	p. 50	p. 58	p. 67
Practice B	p. 9	p. 17	p. 25	p. 33	p. 42	p. 51	p. 59	p. 68
Practice C	p. 10	p. 18	p. 26	p. 34	p. 43	p. 52	p. 60	p. 69
Study Guide	p. 11	p. 19	p. 27	p. 35	p. 44	p. 53	p. 61	p. 70
Real-World Problem Solving					p. 46		p. 63	
Challenge Practice	p. 13	p. 21	p. 29	p. 37	p. 47	p. 55	p. 64	p. 72

REVIEW AND PROJECT MATERIALS

Chapter Review Games and Activities	p. 73
Real-Life Project with Rubric	p. 74
Cooperative Project with Rubric	p. 76
Independent Project with Rubric	p. 78
Cumulative Practice	p. 80
Resource Book Answers	p. A1

Contents

Descriptions of Resources

This Chapter Resource Book is organized by lessons within the chapter in order to make your planning easier. The following materials are provided:

Tips for New Teachers These teaching notes provide both new and experienced teachers with useful teaching tips for each lesson, including tips about common errors and inclusion.

Parents as Partners This guide helps parents contribute to student success by providing an overview of the chapter along with questions and activities for parents and students to work on together.

Lesson Plans and Lesson Plans for Block Scheduling This planning template helps teachers select the materials they will use to teach each lesson from among the variety of materials available for the lesson. The block-scheduling version provides additional information about pacing.

Activity Masters These blackline masters make it easier for students to record their work on selected activities in the Student Edition, or they provide alternative activities for selected answers.

Technology Activities with Keystrokes Keystrokes for various models of calculators are provided for each Technology Activity in the Student Edition where appropriate, along with alternative Technology Activities for selected lessons.

Practice A, B, and C These exercises offer additional practice for the material in each lesson, including application problems. There are three levels of practice for each lesson: A (basic), B (average), and C (advanced).

Study Guide These two pages provide additional instruction, worked-out examples, and practice exercises covering the key concepts and vocabulary in each lesson.

Real-World Problem Solving These exercises offer problem-solving activities for the material in selected lessons in a real world context.

Challenge Practice These exercises offer challenging practice on the mathematics of each lesson for advanced students.

Chapter Review Games and Activities This worksheet offers fun practice at the end of the chapter and provides an alternative way to review the chapter content in preparation for the Chapter Test.

Projects with Rubric These projects allow students to delve more deeply into a problem that applies the mathematics of the chapter. Teacher's notes and a 4-point rubric are included. The projects include a real-life project, a cooperative project, and an independent extra credit project.

Cumulative Practice These practice pages help students maintain skills from the current chapter and preceding chapters.

CHAPTER 9

Tips for New Teachers
For use with Chapter 9

Lesson 9.1

COMMON ERROR After students understand that a positive number has both a positive and a negative square root, reinforce the concept that the radical sign represents only the *nonnegative* square root.

TEACHING TIP It would be beneficial to have students memorize a number of perfect squares and their corresponding square roots.

TEACHING TIP In Example 3 on page 454, reinforce the fact that the square root of a non-perfect square is a non-terminating, non-repeating decimal, and these square roots are not rational numbers.

INCLUSION Some students may require additional practice with the concepts in Example 4 on page 454. Be prepared to provide additional examples, and also be prepared to do some quick assessment on whether your students have grasped the concepts.

COMMON ERROR Before teaching Example 5 on page 455, reinforce the idea that there are two solutions to an equation in which the variable is squared. Have students let the exponent *2* remind them that there are *2* solutions. However, word problems, such as Example 5, often require that the negative square root be discarded as it has no meaning in the real-world context of the problem. This principle will need to be reinforced constantly.

Lesson 9.2

TEACHING TIP When teaching the Product Property of Square Roots on page 458, it may be helpful to compare the similarity between $\sqrt{ab} = \sqrt{a} \cdot \sqrt{b}$ and $(ab)^2 = a^2 b^2$.

TEACHING TIP Stress the two requirements listed on page 458 for a radical expression to be in simplest form. You may want to write these requirements on a poster for display in your room.

TEACHING TIP When teaching Example 1 on page 458, be sure to refer to the Study Strategy in the left margin. Emphasize to students that it is generally easier to simplify the radical in one step by factoring out the *greatest* perfect square factor. The correct answer can still be obtained in more than one step, but it will require students to be diligent in checking further for more perfect square factors.

Lesson 9.3

INCLUSION It will be beneficial to have completed Concept Activity 9.3 on page 464 prior to beginning Lesson 9.3. This activity will give students a good concrete understanding of the Pythagorean theorem.

COMMON ERROR Students must remember that c represents the greatest of the 3 lengths in $a^2 + b^2 = c^2$, which means c always represents the hypotenuse. Compare Example 1 on page 465 with Example 2 on page 466, stressing this fact about c.

TEACHING TIP On page 466, take sufficient time to help students understand the difference between the Pythagorean theorem and the converse of the Pythagorean theorem.

Lesson 9.4

TEACHING TIP Discuss in detail the Venn diagram on page 470. Emphasize that the set of rational numbers and the set of irrational numbers are mutually exclusive.

TEACHING TIP Point out the Study Strategy next to Example 3 on page 471. Remind students that fractions can be rewritten as decimal numbers by dividing their numerators by their denominators. Therefore, an easy way to order a list of real numbers is to convert all of them to decimal numbers.

Tips for New Teachers
For use with Chapter 9

Lesson 9.5

TEACHING TIP Be sure students understand the Reading Algebra margin note on page 476 so that they don't confuse subscripts with exponents or otherwise try to compute with them. If you have two students in class with the same first names, you might point out how it is necessary to use their last names to distinguish them. In this case, last names function just like subscripts.

TEACHING TIP Stress to students that in both the midpoint and distance formulas x_2 and y_2 must come from the same ordered pair, as must x_1 and y_1.

COMMON ERROR When calculating slope, some students end up calculating the reciprocal instead. Stress that the difference of the *y*-coordinates (the rise) goes in the numerator, and the difference of the *x*-coordinates (the run) goes in the denominator. Point out to students that they must *rise* before they can *run*.

Lesson 9.6

TEACHING TIP The following mnemonic device can help students distinguish between 45°-45°-90° triangles and 30°-60°-90° triangles. There are 2 angles of equal measure in a 45°-45°-90° triangle, so the hypotenuse equals $\sqrt{2}$ times either leg. There are 3 different angle measures in a 30°-60°-90° triangle, so the hypotenuse equals $\sqrt{3}$ times the shorter leg.

Lesson 9.7

TEACHING TIP Be sure students understand the meaning of the words *adjacent* and *opposite*. If students are seated at tables, ask them who is seated adjacent to them and who is seated opposite from them.

COMMON ERROR The hypotenuse is neither the adjacent nor opposite side. Adjacent and opposite always refer to the *legs* of a right triangle.

Lesson 9.8

TEACHING TIP Although the concepts of sine and cosine are similar to the concept of tangent, the introduction of the former two often causes confusion for students as to which are which. Suggest the mnemonic device SOHCAHTOA, where

$$\text{Sine} = \frac{\text{Opposite}}{\text{Hypotenuse}}$$

$$\text{Cosine} = \frac{\text{Adjacent}}{\text{Hyponenuse}}, \text{ and}$$

$$\text{Tangent} = \frac{\text{Opposite}}{\text{Adjacent}}.$$

CHAPTER 9

Name _____ Date _____

Parents as Partners
For use with Chapter 9

Chapter Overview One way you can help your student succeed in Chapter 9 is by discussing the lesson goals in the chart below. When a lesson is completed, ask your student the following questions. "What were the goals of the lesson? What new words and formulas did you learn? How can you apply the ideas of the lesson to your life?"

Lesson Title	Lesson Goals	Key Applications
9.1: Square Roots	Find and approximate square roots of numbers.	• Human Chess • Photography • Running
9.2: Simplifying Square Roots	Simplify radical expressions.	• Accident Investigation • Dropped Object • Visual Horizon
9.3: The Pythagorean Theorem	Use the Pythagorean theorem to solve problems.	• Bridges • Television Screen • Synchronized Swimming
9.4: Real Numbers	Compare and order real numbers.	• Fences • Pendulums • Sailing
9.5: The Distance and Midpoint Formulas	Use the distance, midpoint, and slope formulas.	• Bird Watching • Maps • Urban Forest
9.6: Special Right Triangles	Use special right triangles to solve problems.	• Gymnastics • Speakers • Softball
9.7: The Tangent Ratio	Use tangent to find side lengths of right triangles.	• Lunar Formations • Swimming • World's Largest Coffeepot
9.8: The Sine and Cosine Ratios	Use sine and cosine to find triangle side lengths.	• Ancient Sundial • Radio Tower • Ladders

Notetaking Strategies

Using a Concept Map is the strategy featured in Chapter 9 (see page 452). Encourage your student to draw a diagram called a concept map to show connections among key ideas. By connecting the ideas this way on paper, your student is also organizing the ideas and relationships mentally. Your student will be more likely to remember the concepts and how they relate. Your student may want to leave room around his/her concept map in case he/she wants to add additional connections at a later time.

CHAPTER 9 Continued

Parents as Partners

For use with Chapter 9

Name _____ Date _____

Key Ideas Your student can demonstrate understanding of key concepts by working through the following exercises with you.

Lesson	Exercise
9.1	Evaluate the expression when $a = 64$ and $b = 15$. (a) $\sqrt{a-b}$ (b) $\sqrt{a+b+2}$ (c) $\sqrt{b^2 - (a-8)}$
9.2	Simplify the expression. (a) $\sqrt{50xy^2}$ (b) $\sqrt{\dfrac{8a}{36b^2}}$
9.3	Determine whether the triangle with the given side lengths is a right triangle. (a) 24, 45, 51 (b) 17, 19, 25
9.4	Tell whether the number is *rational* or *irrational*. (a) $\sqrt{\dfrac{7}{16}}$ (b) $-\sqrt{144}$ (c) $6.\overline{30}$
9.5	Find the distance between the points. Write your answer in simplest form. (a) (3, 5), (7, 10) (b) (1, −2), (7, 2)
9.6	Find the unknown lengths. Write your answers in simplest form. (a) triangle with 45°-45° angles, hypotenuse $8\sqrt{2}$ ft, legs x (b) triangle with 60°-30° angles, side opposite 60° is $7\sqrt{3}$ m, x and y unknown
9.7	Approximate (a) tan 57° and (b) tan 24° to four decimal places.
9.8	Approximate (a) sin 68° and (b) cos 31° to four decimal places.

Home Involvement Activity

Directions: Use straws or uncooked spaghetti to form triangles using only whole number side lengths. Then use the Pythagorean theorem to determine whether your triangles are right triangles.

Answers

9.1: (a) 7 (b) 9 (c) 13 9.2: (a) $5y\sqrt{2x}$ (b) $\dfrac{\sqrt{2a}}{3b}$ 9.3: (a) yes (b) no 9.4: (a) irrational (b) rational (c) rational 9.5: (a) $\sqrt{41}$ (b) $2\sqrt{13}$ 9.6: (a) $x = 8$ ft (b) $x = 7$ m; $y = 14$ m 9.7: (a) 1.5399 (b) 0.4452 9.8: (a) 0.9272 (b) 0.8572

LESSON 9.1

Teacher's Name _____ Class _____ Room _____ Date _____

Lesson Plan

2-day lesson (See *Pacing and Assignment Guide*, TE page 450A)
For use with pages 453–457

GOAL Find and approximate square roots of numbers.

State/Local Objectives _____

_____

✓ **Check the items you wish to use for this lesson.**

STARTING OPTIONS
____ Warm-Up: Transparencies

TEACHING OPTIONS
____ Notetaking Guide
____ Examples: Day 1: 1, 3, SE pages 453–454; Day 2: 2, 4–5, SE pages 454–455
____ Extra Examples: TE pages 454–455
____ Checkpoint Exercises: Day 1: 1–4, SE page 453; Day 2: 5–8, SE pages 454–455
____ Concept Check: TE page 455
____ Guided Practice Exercises: Day 1: 1–6, SE page 455; Day 2: 7–15, SE page 455
____ Technology Activity: CRB page 7

APPLY/HOMEWORK

Homework Assignment
____ Basic: Day 1: EP p. 803 Exs. 1–8; pp. 456–457 Exs. 16–24, 33–38, 54–57, 60–62, 70–73
   Day 2: pp. 456–457 Exs. 25–30, 41–50, 63–65, 74–83
____ Average: Day 1: pp. 456–457 Exs. 18–24, 33–40, 54–57, 70–77
   Day 2: pp. 456–457 Exs. 27–32, 41–53, 58–62, 66–68, 78–83
____ Advanced: Day 1: pp. 456–457 Exs. 20–24, 37–40, 54–57, 72–75
   Day 2: pp. 456–457 Exs. 29–32, 43–53, 58, 59, 63–69*, 78–83

Reteaching the Lesson
____ Practice: CRB pages 8–10 (Level A, Level B, Level C); Practice Workbook
____ Study Guide: CRB pages 11–12; Spanish Study Guide

Extending the Lesson
____ Challenge: SE page 457; CRB page 13

ASSESSMENT OPTIONS
____ Daily Quiz (9.1): TE page 457 or Transparencies
____ Standardized Test Practice: SE page 457

Notes _____

_____

_____

LESSON 9.1

Teacher's Name _____ Class _____ Room _____ Date _____

Lesson Plan for Block Scheduling

1-block lesson (See *Pacing and Assignment Guide*, TE page 450A)
For use with pages 453–457

GOAL Find and approximate square roots of numbers.

State/Local Objectives _____

_____

✓ **Check the items you wish to use for this lesson.**

STARTING OPTIONS
____ Warm-Up: Transparencies

TEACHING OPTIONS
____ Notetaking Guide
____ Examples: 1–5, SE pages 453–455
____ Extra Examples: TE pages 454–455
____ Checkpoint Exercises: 1–8, SE pages 453–455
____ Concept Check: TE page 455
____ Guided Practice Exercises: 1–15, SE page 455
____ Technology Activity: CRB page 7

Chapter Pacing Guide	
Day	Lesson
1	**9.1**
2	9.2
3	9.3; 9.4
4	9.5; 9.6 (begin)
5	9.6 (end); 9.7
6	9.8
7	Ch. 9 Review and Projects

APPLY/HOMEWORK

Homework Assignment
____ Block Schedule: pp. 456–457 Exs. 18–24, 27–62, 66–68, 70–83

Reteaching the Lesson
____ Practice: CRB pages 8–10 (Level A, Level B, Level C); Practice Workbook
____ Study Guide: CRB pages 11–12; Spanish Study Guide

Extending the Lesson
____ Challenge: SE page 457; CRB page 13

ASSESSMENT OPTIONS
____ Daily Quiz (9.1): TE page 457 or Transparencies
____ Standardized Test Practice: SE page 457

Notes _____

_____

_____

_____

LESSON 9.1 Technology Activity
For use with pages 453–457

GOAL Use a calculator to evaluate radical expressions.

EXAMPLE Evaluate the expression when $a = 5$ and $b = 3.2$. Round to the nearest hundredth if necessary.

a. $\sqrt{\dfrac{b}{a}}$ b. $\sqrt{\dfrac{a^2 - b^2}{2}}$

Solution

a. Substitute 5 for a and 3.2 for b in the original expression. Then use the following keystrokes to evaluate the expression.

Keystrokes: 2nd [√] 3.2 ÷ 5) =

Answer: $\sqrt{\dfrac{3.2}{5}} = 0.8$

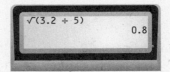

b. Substitute 5 for a and 3.2 for b in the original expression. Then use the following keystrokes to evaluate the expression.

Keystrokes: 2nd [√] (5 x² − 3.2 x²) ÷ 2) =

Answer: $\sqrt{\dfrac{5^2 - 3.2^2}{2}} \approx 2.72$

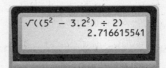

Technology Tip
Your calculator performs the established order of operations. So the use of parentheses is very important when evaluating expressions on a calculator.

DRAW CONCLUSIONS Approximate the square root to the nearest hundredth.

1. $\sqrt{52}$ 2. $\sqrt{112}$ 3. $\sqrt{500}$
4. $\sqrt{\dfrac{8}{15}}$ 5. $\sqrt{27 + 3.6}$ 6. $\sqrt{41 - 13.5}$

Use a calculator to evaluate the expression when $a = 4.1$ and $b = 8.4$. Round to the nearest hundredth.

7. $\sqrt{ab}$ 8. $\sqrt{a^2 + b}$ 9. $\sqrt{\dfrac{a}{b}}$
10. $\sqrt{b - a}$ 11. $\sqrt{\dfrac{b^2 - 3}{a}}$ 12. $\sqrt{\dfrac{a^2 - b}{5}}$

LESSON 9.1 Practice A

For use with pages 453–457

Find the square roots of the number.

1. 49
2. 64
3. 100
4. 196
5. 400
6. 2500

Approximate the square root to the nearest integer.

7. $\sqrt{23}$
8. $-\sqrt{95}$
9. $\sqrt{77}$
10. $-\sqrt{125}$
11. $-\sqrt{32}$
12. $\sqrt{10.2}$

Use a calculator to approximate the square root. Round to the nearest tenth.

13. $\sqrt{2}$
14. $-\sqrt{8}$
15. $\sqrt{91}$
16. $-\sqrt{53}$
17. $\sqrt{1175}$
18. $\sqrt{14.5}$

Evaluate the expression when $a = 36$ and $b = 9$. Round to the nearest tenth if necessary.

19. $\sqrt{a-b}$
20. $\sqrt{a+b+4}$
21. $-2\sqrt{ab}$

Solve the equation. Round to the nearest tenth if necessary.

22. $x^2 = 36$
23. $y^2 = 169$
24. $196 = n^2$
25. $t^2 = 29$
26. $110 = c^2$
27. $4y^2 = 33$

Solve the equation. Round to the nearest hundredth if necessary.

28. $x^2 + 4 = 29$
29. $120 = 0.5c^2$
30. $3y^2 - 5 = 23$

In Exercises 31–33, use the following information. You are helping build a sandbox for a community park. You want the sandbox to be in the shape of a square with an area of 30 square feet. You have boards with a total length of 24 feet to use as sides.

31. Find the side length of the sandbox to the nearest tenth of a foot.
32. Find the perimeter of the sandbox.
33. Do you have enough boards to make the sides of the sandbox?

LESSON 9.1 Practice B

For use with pages 453–457

Find the square roots of the number.

1. 36
2. 361
3. 729
4. 1089
5. 4900
6. 10,000

Approximate the square root to the nearest integer.

7. $\sqrt{39}$
8. $-\sqrt{85}$
9. $\sqrt{105}$
10. $-\sqrt{136}$
11. $\sqrt{17.4}$
12. $-\sqrt{3.3}$

Use a calculator to approximate the square root. Round to the nearest tenth.

13. $\sqrt{5}$
14. $-\sqrt{12}$
15. $\sqrt{102}$
16. $-\sqrt{74}$
17. $\sqrt{1585}$
18. $\sqrt{27.8}$

Evaluate the expression when $a = 72$ and $b = 8$.

19. $\sqrt{a-b}$
20. $\sqrt{a+b+1}$
21. $-4\sqrt{ab}$

Solve the equation. Round to the nearest tenth if necessary.

22. $x^2 = 64$
23. $y^2 = 324$
24. $225 = n^2$
25. $t^2 = 42$
26. $150 = c^2$
27. $5y^2 = 48$

Solve the equation. Round to the nearest hundredth if necessary.

28. $2x^2 = 32$
29. $90 = 1.5t^2 + 8$
30. $5n^2 - 4 = 74$

31. A square ice skating rink has an area of 1849 square feet. What is the perimeter of the rink?

32. A forest ranger is stationed in a 58 foot tall fire tower. The equation for the distance in miles that the ranger can see is $d = \sqrt{1.5h}$, where h is the height in feet above the ground. Find the distance the ranger can see. Round your answer to the nearest tenth.

LESSON 9.1 Practice C
For use with pages 453–457

Approximate the square root to the nearest integer.

1. $\sqrt{46}$
2. $-\sqrt{99}$
3. $\sqrt{115}$
4. $-\sqrt{140}$
5. $-\sqrt{7.3}$
6. $\sqrt{8.63}$

Use a calculator to approximate the square root. Round to the nearest tenth.

7. $\sqrt{11}$
8. $-\sqrt{29}$
9. $\sqrt{107}$
10. $-\sqrt{215}$
11. $\sqrt{1234}$
12. $\sqrt{9.99}$

Evaluate the expression when $a = 156$ and $b = 12$.

13. $\sqrt{a-b}$
14. $\sqrt{a+b+1}$
15. $-5\sqrt{a-(b^2-4)}$

Solve the equation. Round to the nearest tenth if necessary.

16. $x^2 = 121$
17. $y^2 = 441$
18. $2704 = n^2$
19. $t^2 = 59$
20. $1100 = 5c^2$
21. $9y^2 = 633$

Solve the equation. Round to the nearest hundredth if necessary.

22. $x^2 + 8 = 31$
23. $144 = 0.5m^2$
24. $39 = 6y^2 - 9$

25. Write an equation that has exactly two solutions, -0.13 and 0.13.

26. A school is building a playground on a square plot of land that has an area of 650 square meters. The school wants to build a fence around the plot of land and currently has 60 meters of fencing.

$A = 650 \text{ m}^2$

a. Find the length of a side of the plot of land to the nearest tenth of a yard.

b. Use part (a) to approximate the perimeter of the plot of land.

c. Does the school have enough fencing to build a fence around the plot of land? If not, about how much more do they need?

LESSON 9.1

Name _____ Date _____

Study Guide
For use with pages 453–457

GOAL Find and approximate square roots of numbers.

> **VOCABULARY**
>
> A **square root** of a number n is a number m such that $m^2 = n$.
> A **perfect square** is a number that is the square of an integer.
> A **radical expression** is an expression that involves a radical sign.

EXAMPLE 1 Finding a Square Root

The area of a square is 225 square feet. Find the side length of the square.

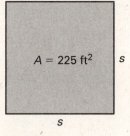

$A = 225$ ft^2 s

s

Solution

The side length of the square is the positive square root of its area.

$\sqrt{225} = 15$ because $15^2 = 225$.

Answer: The side length of the square is 15 feet.

Exercises for Example 1

Find the square roots of the number.

1. 25   **2.** 9   **3.** 100   **4.** 121

EXAMPLE 2 Approximating a Square Root

Approximate $-\sqrt{60}$ to the nearest integer.

$49 < 60 < 64$	Identify perfect squares closest to 60.
$-\sqrt{49} > -\sqrt{60} > -\sqrt{64}$	Take negative square root of each number. Reverse inequality symbols.
$-7 > -\sqrt{60} > -8$	Evaluate square root of each perfect square.

Answer: Because 60 is closer to 64 than to 49, $-\sqrt{60}$ is closer to -8 than to -7. So, to the nearest integer, $-\sqrt{60} \approx -8$.

Exercises for Example 2

Approximate the square root to the nearest integer.

5. $\sqrt{14}$ **6.** $\sqrt{26}$ **7.** $-\sqrt{56.8}$ **8.** $\sqrt{110.21}$

LESSON 9.1 Continued

Study Guide
For use with pages 453–457

EXAMPLE 3 Using a Calculator

Use a calculator to approximate $-\sqrt{389}$. Round to the nearest tenth.

Keystrokes

(−) 2nd [√] 389) =

Answer: $-\sqrt{389} \approx -19.7$

Display

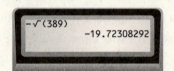

Exercises for Example 3

Use a calculator to approximate the square root. Round to the nearest tenth.

9. $\sqrt{15}$ 10. $-\sqrt{95}$ 11. $-\sqrt{3.7}$ 12. $\sqrt{74.89}$

EXAMPLE 4 Evaluating a Radical Expression

Evaluate $\sqrt{3x^3 + y^2 - 2x}$ when $x = 5$ and $y = -9$.

Solution

$3\sqrt{x^3 + y^2 - 2x} = 3\sqrt{5^3 + (-9)^2 - 2(5)}$ Substitute 5 for x and -9 for y.
$= 3\sqrt{196}$ Evaluate expression inside radical symbol.
$= 3 \cdot 14$ Evaluate square root.
$= 42$ Multiply.

EXAMPLE 5 Solving an Equation Using Square Roots

Solve $7x^2 - 5 = 562$.

Solution

$7x^2 - 5 = 562$ Write original equation.
$7x^2 = 567$ Add 5 to each side.
$x^2 = 81$ Divide each side by 7.
$x = \pm\sqrt{81}$ Use definition of square root.
$x = \pm 9$ Evaluate square root.

Answer: The solutions are 9 and -9.

Exercises for Examples 4 and 5

In Exercises 13–15, evaluate the expression when $x = 10$ and $y = 2$.

13. $\sqrt{5x^2 - 50y}$ 14. $-\sqrt{3x^2 - 22y}$ 15. $-\sqrt{8x + 2y^5}$

16. Solve $3x^2 - 11 = 64$.

LESSON 9.1 Challenge Practice

For use with pages 453–457

Evaluate the expression when $a = 28$ and $b = 53$.

1. $-6\sqrt{a^2 + b + 4}$
2. $4a\sqrt{b - a}$
3. $-8\sqrt{5a + 2b - 21}$

Solve the equation. Round to the nearest hundredth if necessary.

4. $\frac{1}{4}x^2 + 15 = 47$
5. $3.2r^2 - 4.1 = 2.3$
6. $8(m^2 - 4) + 9 = 13$

Copy and complete the statement using <, >, or =.

7. $\sqrt{51}$ __?__ $7\frac{1}{5}$
8. $\sqrt{20.25}$ __?__ 4.5
9. -8.28 __?__ $-\sqrt{68}$

10. Solve $(3 - x)^2 - 4 = 45$. Describe the steps you use.

LESSON Teacher's Name _____ Class _____ Room _____ Date _____
9.2
Lesson Plan
2-day lesson (See *Pacing and Assignment Guide*, TE page 450A)
For use with pages 458–461

GOAL Simplify radical expressions.

State/Local Objectives _____
_____

✓ **Check the items you wish to use for this lesson.**

STARTING OPTIONS
____ Homework Check (9.1): TE page 456; Answer Transparencies
____ Homework Quiz (9.1): TE page 457; Transparencies
____ Warm-Up: Transparencies

TEACHING OPTIONS
____ Notetaking Guide
____ Examples: Day 1: 1, SE page 458; Day 2: 2–4, SE pages 458–459
____ Extra Examples: TE page 459
____ Checkpoint Exercises: Day 1: 1, SE page 459; Day 2: 2–4, SE page 459
____ Concept Check: TE page 459
____ Guided Practice Exercises: Day 1: 1–4, 7–8, SE page 460; Day 2: 5–6, SE page 460

APPLY/HOMEWORK
Homework Assignment
____ Basic: Day 1: pp. 460–461 Exs. 9–14, 31, 34–41
  Day 2: pp. 460–461 Exs. 15–26, 42–47
____ Average: Day 1: pp. 460–461 Exs. 9–14, 29–32, 34–41
  Day 2: pp. 460–461 Exs. 15–28, 42–47
____ Advanced: Day 1: pp. 460–461 Exs. 11–14, 28–33*, 44–46
  Day 2: pp. 460–461 Exs. 15–27, 36–43, 47

Reteaching the Lesson
____ Practice: CRB pages 16–18 (Level A, Level B, Level C); Practice Workbook
____ Study Guide: CRB pages 19–20; Spanish Study Guide

Extending the Lesson
____ Challenge: SE page 461; CRB page 21

ASSESSMENT OPTIONS
____ Daily Quiz (9.2): TE page 461 or Transparencies
____ Standardized Test Practice: SE page 461

Notes _____
_____

LESSON 9.2

Teacher's Name _____ Class _____ Room _____ Date _____

Lesson Plan for Block Scheduling

1-block lesson (See *Pacing and Assignment Guide*, TE page 450A)
For use with pages 458–461

GOAL   Simplify radical expressions.

State/Local Objectives _____

_____

✓ **Check the items you wish to use for this lesson.**

STARTING OPTIONS

____ Homework Check (9.1): TE page 456; Answer Transparencies
____ Homework Quiz (9.1): TE page 457; Transparencies
____ Warm-Up: Transparencies

TEACHING OPTIONS

____ Notetaking Guide
____ Examples: 1–4, SE pages 458–459
____ Extra Examples: TE page 459
____ Checkpoint Exercises: 1–4, SE page 459
____ Concept Check: TE page 459
____ Guided Practice Exercises: 1–8, SE page 460

Chapter Pacing Guide	
Day	Lesson
1	9.1
2	**9.2**
3	9.3; 9.4
4	9.5; 9.6 (begin)
5	9.6 (end); 9.7
6	9.8
7	Ch. 9 Review and Projects

APPLY/HOMEWORK

Homework Assignment

____ Block Schedule: pp. 460–461 Exs. 9–32, 34–47

Reteaching the Lesson

____ Practice: CRB pages 16–18 (Level A, Level B, Level C); Practice Workbook
____ Study Guide: CRB pages 19–20; Spanish Study Guide

Extending the Lesson

____ Challenge: SE page 461; CRB page 21

ASSESSMENT OPTIONS

____ Daily Quiz (9.2): TE page 461 or Transparencies
____ Standardized Test Practice: SE page 461

Notes _____

_____

_____

_____

Copyright © McDougal Littell/Houghton Mifflin Company
All rights reserved.

Pre-Algebra   **15**
Chapter 9   Resource Book

LESSON 9.2 Practice A

For use with pages 458–461

Simplify the expression.

1. $\sqrt{20}$
2. $\sqrt{150}$
3. $\sqrt{252}$
4. $\sqrt{175}$
5. $\sqrt{432y}$
6. $\sqrt{48c^2}$
7. $\sqrt{\dfrac{17}{49}}$
8. $\sqrt{\dfrac{23}{81}}$
9. $\sqrt{\dfrac{45}{144}}$
10. $\sqrt{\dfrac{84}{169}}$
11. $\sqrt{\dfrac{p}{100}}$
12. $\sqrt{\dfrac{32x^2}{25}}$

13. A square has an area of 500 square units. Find the length of a side of the square as a radical expression in simplest form.

Simplify the expression.

14. $\sqrt{60b^2c}$
15. $\sqrt{2400k^2\ell^2}$
16. $\sqrt{120h^2j}$
17. $\sqrt{\dfrac{600x}{9y^2}}$
18. $\sqrt{\dfrac{7m^2}{225n^2}}$
19. $\sqrt{\dfrac{65y}{16z^2}}$

20. After a car accident on a snow-packed road, an investigator measures the length ℓ (in feet) of a car's skid marks. The expression $\sqrt{12\ell}$ gives the car's speed in miles per hour at the time the brakes were applied.

    a. Write the expression in simplest form.

    b. The skid marks are 170 feet long. Use the simplified expression to approximate the car's speed when the brakes were applied.

21. You stand on a bridge and drop a pebble into the water below from a height of 75 feet. You can use the expression $\sqrt{\dfrac{75}{16}}$ to find the time in seconds that it takes the pebble to hit the water. Write the expression in simplest form. Then approximate the value of the expression to the nearest second.

LESSON 9.2 Practice B

For use with pages 458–461

Simplify the expression.

1. $\sqrt{54}$
2. $\sqrt{112}$
3. $\sqrt{176}$
4. $\sqrt{180}$
5. $\sqrt{117f}$
6. $\sqrt{432y^2}$
7. $\sqrt{\dfrac{120}{121}}$
8. $\sqrt{\dfrac{75}{225}}$
9. $\sqrt{\dfrac{202}{256}}$
10. $\sqrt{\dfrac{320}{441}}$
11. $\sqrt{\dfrac{21v^2}{324}}$
12. $\sqrt{\dfrac{94t}{196}}$

13. A square has an area of 700 square units. Find the length of a side of the square as a radical expression in simplest form.

Simplify the expression.

14. $\sqrt{171cd^2}$
15. $\sqrt{152m^2n}$
16. $\sqrt{126x^2y^2}$
17. $\sqrt{\dfrac{23w^2}{49}}$
18. $\sqrt{\dfrac{45rt^2}{144}}$
19. $\sqrt{\dfrac{76p^2q^2}{81}}$

20. After a car accident on a dry asphalt road, an investigator measures the length ℓ (in feet) of a car's skid marks. The expression $\sqrt{18\ell}$ gives the car's speed in miles per hour at the time the brakes were applied.

    a. Write the expression in simplest form.

    b. The skid marks are 140 feet long. Use the simplified expression to approximate the car's speed when the brakes were applied.

21. You drop a stick from the top of a building that is 245 feet high. You can use the expression $\sqrt{\dfrac{245}{16}}$ to find the time in seconds that it takes the stick to hit the ground. Write the expression in simplest form. Then approximate the value of the expression to the nearest second.

LESSON 9.2 Practice C

For use with pages 458–461

Simplify the expression.

1. $\sqrt{350}$
2. $\sqrt{735}$
3. $\sqrt{448}$
4. $\sqrt{108}$
5. $\sqrt{184x^2y^3}$
6. $\sqrt{27m^3n}$
7. $\sqrt{\dfrac{62}{361}}$
8. $\sqrt{\dfrac{68}{196}}$
9. $\sqrt{\dfrac{147}{49}}$
10. $\sqrt{\dfrac{135}{256}}$
11. $\sqrt{\dfrac{96t}{289r^2}}$
12. $\sqrt{\dfrac{120b^3}{169a^2}}$

13. A square has an area of 50,000 square units. Find the length of a side of the square as a radical expression in simplest form.

Simplify the expression.

14. $\sqrt{189xy^2z}$
15. $\sqrt{116m^2n^2p^2}$
16. $\sqrt{156a^3bc^2}$
17. $\sqrt{\dfrac{99f}{144d^2}}$
18. $\sqrt{\dfrac{98}{441x^2}}$
19. $\sqrt{\dfrac{176w^2}{484}}$

20. You drop a pebble from a cliff that is 1380 feet high. You can use the expression $\sqrt{\dfrac{1380}{16}}$ to find the time in seconds that it takes the pebble to hit the ground. Write the expression in simplest form. Then approximate the value to the nearest second.

21. The visual horizon is the distance you can see before your line of sight is blocked by Earth's surface. If you are standing on a pier by the ocean, the visual horizon in miles can be approximated by the expression $1.23\sqrt{h}$, where h is the vertical distance in feet from your eye to the ground. Your eyes are 24 feet from the ground. Write the expression for your visual horizon in simplest form. Then approximate the value of the expression to the nearest mile.

22. Predict the next three numbers in the pattern:

    $2, 2\sqrt{2}, 2\sqrt{3}, 4, 2\sqrt{5}, 2\sqrt{6}, \ldots$

LESSON 9.2

Name _____ **Date** _____

Study Guide
For use with pages 458–461

GOAL Simplify radical expressions.

> **VOCABULARY**
>
> A radical expression is in **simplest form** when:
> - No factor of the expression under the radical sign has any perfect square factor other than 1.
> - There are no fractions under the radical sign, and no radical sign in the denominator of any fraction.
>
> **Product Property of Square Roots**
> $\sqrt{ab} = \sqrt{a} \cdot \sqrt{b}$, where $a \geq 0$ and $b \geq 0$
>
> **Quotient Property of Square Roots**
> $\sqrt{\dfrac{a}{b}} = \dfrac{\sqrt{a}}{\sqrt{b}}$ where $a \geq 0$ and $b > 0$

EXAMPLE 1 Simplifying a Radical Expression

$\sqrt{112} = \sqrt{16 \cdot 7}$ Factor using greatest perfect square factor.

$\quad\quad\ = \sqrt{16} \cdot \sqrt{7}$ Product property of square roots

$\quad\quad\ = 4\sqrt{7}$ Simplify.

EXAMPLE 2 Simplifying a Variable Expression

$\sqrt{72a^2} = \sqrt{36 \cdot 2 \cdot a^2}$ Factor using greatest perfect square factor.

$\quad\quad\ = \sqrt{36} \cdot \sqrt{2} \cdot \sqrt{a^2}$ Product property of square roots

$\quad\quad\ = 6 \cdot \sqrt{2} \cdot a$ Simplify.

$\quad\quad\ = 6a\sqrt{2}$ Commutative property

Exercises for Examples 1 and 2

Simplify the expression.

1. $\sqrt{24}$
2. $\sqrt{300}$
3. $\sqrt{275}$
4. $-\sqrt{1000}$
5. $\sqrt{92y^2}$
6. $\sqrt{12x}$
7. $-\sqrt{18t}$
8. $-\sqrt{27k^2}$

LESSON 9.2 Continued

Study Guide
For use with pages 458–461

EXAMPLE 3 Simplifying a Radical Expression

$\sqrt{\dfrac{45}{169}} = \dfrac{\sqrt{45}}{\sqrt{169}}$ Quotient property of square roots

$\phantom{\sqrt{\dfrac{45}{169}}} = \dfrac{3\sqrt{5}}{13}$ Simplify.

Exercises for Example 3

Simplify the expression.

9. $-\sqrt{\dfrac{40}{81x^2}}$ 10. $\sqrt{\dfrac{108}{10{,}000t^2}}$ 11. $-\sqrt{\dfrac{9h}{16}}$ 12. $-\sqrt{\dfrac{50a^2}{289}}$

EXAMPLE 4 Using Radical Expressions

The expression $\sqrt{\dfrac{2h}{g}}$ gives the time t (in seconds) that it takes an object to hit the ground when dropped from a height h and with acceleration due to gravity g. Use the expression to approximate the amount of time it takes a ball dropped from a height of 50 meters to hit the surface of the planet Mars, where the acceleration due to gravity is about 3.7 meters per second per second.

Solution

$\sqrt{\dfrac{2h}{g}} = \sqrt{\dfrac{2 \cdot 50}{3.7}}$ Substitute 50 for h and 3.7 for g.

$\phantom{\sqrt{\dfrac{2h}{g}}} = \sqrt{\dfrac{100}{3.7}}$ Multiply.

$\phantom{\sqrt{\dfrac{2h}{g}}} = \dfrac{\sqrt{100}}{\sqrt{3.7}}$ Quotient property of square roots

$\phantom{\sqrt{\dfrac{2h}{g}}} = \dfrac{10}{\sqrt{3.7}}$ Evaluate square root of 100.

$\phantom{\sqrt{\dfrac{2h}{g}}} \approx 5$ Approximate using a calculator.

Answer: It takes the ball about 5 seconds to hit the ground.

Exercise for Example 4

13. Use the information in Example 4 to find the approximate amount of time it takes a ball to hit the ground when dropped from a height of 50 meters on Earth, where the acceleration due to gravity is about 9.8 meters per second per second.

LESSON 9.2 Challenge Practice

For use with pages 458–461

Simplify the expression.

1. $\sqrt{\dfrac{28}{81x^2}}$

2. $\sqrt{\dfrac{75a}{64b^2}}$

3. $4\sqrt{\dfrac{24m^2n}{9p^2}}$

4. $\sqrt{8x^3}$

5. $\sqrt{12y^5}$

6. $\sqrt{\dfrac{4m^3}{9n^4}}$

7. $\sqrt{4x^2} \cdot \sqrt{52y^2}$

8. $\dfrac{3}{2m}\sqrt{8m^2}$

9. $3r\sqrt{\dfrac{25}{r^2}}$

10. Predict the next three numbers in the pattern:
    $\sqrt{3}, \sqrt{6}, 3, 2\sqrt{3}, \sqrt{15}, 3\sqrt{2}, \sqrt{21}, \ldots$

LESSON 9.3

Teacher's Name _____ Class _____ Room _____ Date _____

Lesson Plan

1-day lesson (See *Pacing and Assignment Guide*, TE page 450A)
For use with pages 464–469

GOAL Use the Pythagorean theorem to solve problems.

State/Local Objectives _____

_____

✓ **Check the items you wish to use for this lesson.**

STARTING OPTIONS
____ Homework Check (9.2): TE page 460; Answer Transparencies
____ Homework Quiz (9.2): TE page 461; Transparencies
____ Warm-Up: Transparencies

TEACHING OPTIONS
____ Notetaking Guide
____ Concept Activity: SE page 464
____ Examples: 1–3, SE pages 465–466
____ Extra Examples: TE page 466
____ Checkpoint Exercises: 1–3, SE page 466
____ Concept Check: TE page 466
____ Guided Practice Exercises: 1–6, SE page 467

APPLY/HOMEWORK
Homework Assignment
____ Basic: pp. 467–469 Exs. 7–16, 19, 23, 25–27, 31, 37–43
____ Average: pp. 467–469 Exs. 9–12, 16–24, 28–34, 37–44
____ Advanced: pp. 467–469 Exs. 9–12, 16–21, 24, 28–38*, 41–44

Reteaching the Lesson
____ Practice: CRB pages 24–26 (Level A, Level B, Level C); Practice Workbook
____ Study Guide: CRB pages 27–28; Spanish Study Guide

Extending the Lesson
____ Challenge: SE page 469; CRB page 29

ASSESSMENT OPTIONS
____ Daily Quiz (9.3): TE page 469 or Transparencies
____ Standardized Test Practice: SE page 469

Notes _____

_____

LESSON
9.3

Teacher's Name _____ Class _____ Room _____ Date _____

Lesson Plan for Block Scheduling

Half-block lesson (See *Pacing and Assignment Guide*, TE page 450A)
For use with pages 464–469

GOAL Use the Pythagorean theorem to solve problems.

State/Local Objectives _____

_____

✓ **Check the items you wish to use for this lesson.**

STARTING OPTIONS

___ Homework Check (9.2): TE page 460; Answer Transparencies
___ Homework Quiz (9.2): TE page 461; Transparencies
___ Warm-Up: Transparencies

TEACHING OPTIONS

___ Notetaking Guide
___ Concept Activity: SE page 464
___ Examples: 1–3, SE pages 465–466
___ Extra Examples: TE page 466
___ Checkpoint Exercises: 1–3, SE page 466
___ Concept Check: TE page 466
___ Guided Practice Exercises: 1–6, SE page 467

Chapter Pacing Guide	
Day	Lesson
1	9.1
2	9.2
3	**9.3**; 9.4
4	9.5; 9.6 (begin)
5	9.6 (end); 9.7
6	9.8
7	Ch. 9 Review and Projects

APPLY/HOMEWORK

Homework Assignment
___ Block Schedule: pp. 467–469 Exs. 9–12, 16–24, 28–34, 37–44 (with 9.4)

Reteaching the Lesson
___ Practice: CRB pages 24–26 (Level A, Level B, Level C); Practice Workbook
___ Study Guide: CRB pages 27–28; Spanish Study Guide

Extending the Lesson
___ Challenge: SE page 469; CRB page 29

ASSESSMENT OPTIONS

___ Daily Quiz (9.3): TE page 469 or Transparencies
___ Standardized Test Practice: SE page 469

Notes _____

_____

_____

_____

LESSON 9.3 Practice A

For use with pages 464–469

Find the unknown length. Write your answer in simplest form.

1.
2.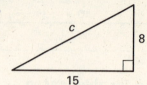
3.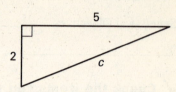
4.
5.
6.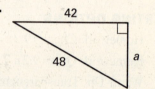

Determine whether the triangle with the given side lengths is a right triangle.

7. 3, 5, 6
8. 5, 12, 13
9. 7, 11, 13
10. 7, 24, 25
11. 10, 50, 51
12. 16, 30, 34

The lengths of two sides of a right triangle are given. Find the length of the third side.

13. $a = 20, c = 29$
14. $a = 9, c = 41$
15. $b = 9, c = 15$
16. $b = 36, c = 45$
17. $a = 28, b = 45$
18. $a = 33, b = 56$

19. A ladder that is 12 feet long is placed against a wall. The bottom of the ladder is 4 feet from the wall. To the nearest foot, how far up the wall does the ladder reach?

    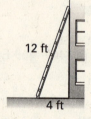

    a. Label the known lengths. Label the unknown length x.

    b. Use the Pythagorean theorem to write an equation you can use to find the value of x.

    c. Solve the equation in part (b) for x.

    d. Round your answer from part (c) to the nearest whole number.

20. A sheet of paper is 8.5 inches by 11 inches. How long is the diagonal of the paper? Round your answer to the nearest tenth of an inch.

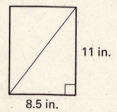

LESSON 9.3 **Practice B**
For use with pages 464–469

Find the unknown length. Write your answer in simplest form.

1.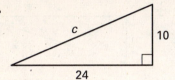
2.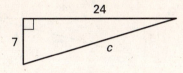
3.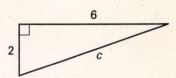
4.
5.
6.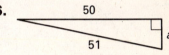

Determine whether the triangle with the given side lengths is a right triangle.

7. 8, 15, 17
8. 20, 21, 28
9. 9, 12, 15
10. 11, 13, 17
11. 5, 64, 65
12. 12, 25, 27

The lengths of two sides of a right triangle are given. Find the length of the third side.

13. $a = 9$, $c = 41$
14. $a = 40$, $c = 58$
15. $b = 56$, $c = 65$
16. $b = 70$, $c = 74$
17. $a = 13$, $b = 84$
18. $a = 16$, $b = 63$

19. A support wire 12 yards long is attached to the top of a utility pole 10 yards tall and is then stretched taut. To the nearest tenth of a yard, find how far from the base of the pole the wire will be attached to the ground.

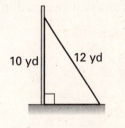

20. An access ramp has a height of 5 feet and a horizontal distance of 60 feet. Find the length ℓ of the ramp to the nearest tenth of a foot.

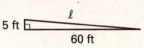

Find the unknown length. Round to the nearest hundredth, if necessary.

21.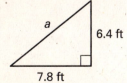
22.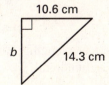
23.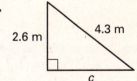

LESSON 9.3 **Practice C**
For use with pages 464–469

Find the unknown length. Write your answer in simplest form.

1.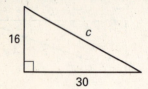
2.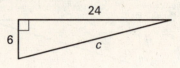
3.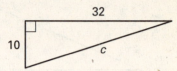
4.
5.
6.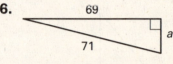

Determine whether the triangle with the given side lengths is a right triangle.

7. 20, 21, 29
8. 5, 15, 16
9. 9, 40, 41
10. 7, 13, 15
11. 12, 73, 74
12. 14, 23, 27

The lengths of two sides of a right triangle are given. Find the length of the third side.

13. $a = 14$, $c = 50$
14. $a = 16$, $c = 34$
15. $b = 35$, $c = 37$
16. $b = 48$, $c = 52$
17. $a = 32$, $b = 60$
18. $a = 28$, $b = 96$

19. A baseball diamond is a square with side lengths of 90 feet. What is the distance from first base to third base? Round to the nearest tenth of a foot.

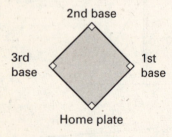

20. You are moving into a new house. The doorway is 78 inches high and 36 inches wide. Can a round table top with a diameter of 84 inches fit through the doorway?

21. The hypotenuse of an isosceles right triangle has a length of 9 meters. Find the leg lengths to the nearest hundredth of a meter.

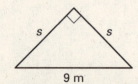

LESSON 9.3

Study Guide
For use with pages 464–469

GOAL Use the Pythagorean theorem to solve problems.

VOCABULARY

In a right triangle, the **hypotenuse** is the side opposite the right angle. The **legs** are the sides that form the right angle. The lengths of the legs and the length of the hypotenuse of a right triangle are related by the **Pythagorean theorem**, which states that for any right triangle, the sum of the squares of the lengths of the legs equals the square of the length of the hypotenuse.

EXAMPLE 1 Finding the Length of a Hypotenuse

City A is 30 miles due north of City B. City C is 40 miles due east of City B. Find the distance c between City A and City C.

Solution

To find the distance between the cities, picture a right triangle connecting the cities.

$a^2 + b^2 = c^2$	Pythagorean theorem
$30^2 + 40^2 = c^2$	Substitute 30 for a and 40 for b.
$900 + 1600 = c^2$	Evaluate powers.
$2500 = c^2$	Add.
$\sqrt{2500} = c$	Take positive square root of each side.
$50 = c$	Evaluate square root.

Answer: The distance between City A and City C is 50 miles.

EXAMPLE 2 Finding the Length of a Leg

Find the unknown length b in simplest form.

Solution

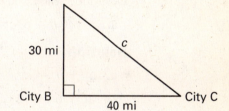

$a^2 + b^2 = c^2$	Pythagorean theorem
$6^2 + b^2 = 14^2$	Substitute.
$36 + b^2 = 196$	Evaluate powers.
$b^2 = 160$	Subtract 36 from each side.
$b = \sqrt{160}$	Take positive square root of each side.
$b = 4\sqrt{10}$	Simplify.

Answer: The unknown length b is $4\sqrt{10}$ units.

LESSON 9.3 Continued

Study Guide
For use with pages 464–469

Exercises for Examples 1 and 2

Find the unknown length. Write your answer in simplest form.

1.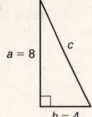

2.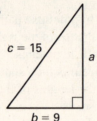

3.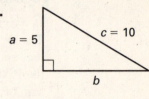

EXAMPLE 3 Identifying Right Triangles

Determine whether the triangle with the given side lengths is a right triangle.

a. $a = 8, b = 6, c = 12$ 　　　**b.** $a = 3, b = 4, c = 5$

Solution

a. $a^2 + b^2 = c^2$

$8^2 + 6^2 \stackrel{?}{=} 12^2$

$64 + 36 \stackrel{?}{=} 144$

$100 \neq 144$

Answer: Not a right triangle

b. $a^2 + b^2 = c^2$

$3^2 + 4^2 \stackrel{?}{=} 5^2$

$9 + 16 \stackrel{?}{=} 25$

$25 = 25$ ✓

Answer: A right triangle

Exercises for Example 3

Determine whether the triangle with the given side lengths is a right triangle.

4. $a = 12, b = 5, c = 13$
5. $a = 3, b = 9, c = 11$
6. $a = 6, b = 11, c = 12$

LESSON 9.3 Challenge Practice

For use with pages 464–469

Find the unknown length. Write your answer in simplest form.

1.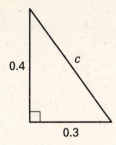

2.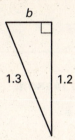

3.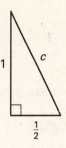

4. Find the length of the diagonal of a rectangle that is 15 centimeters long and 12 centimeters wide. Round your answer to the nearest centimeter.

Find the value of x. Round your answer to the nearest hundredth if necessary.

5.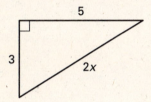

6.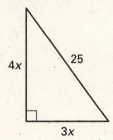

7.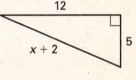

8. Find the lengths of the sides of the triangle whose vertices are at (0, 0), (8, 0), and (0, 6).

9. Find the area of a right triangle whose hypotenuse is 7 inches long and has a leg that is 4 inches long. Round your answer to the nearest inch.

LESSON Teacher's Name _____ Class _____ Room _____ Date _____

9.4 Lesson Plan

1-day lesson (See *Pacing and Assignment Guide*, TE page 450A)
For use with pages 470–474

GOAL   Compare and order real numbers.

State/Local Objectives _____

_____

✓ **Check the items you wish to use for this lesson.**

STARTING OPTIONS

____ Homework Check (9.3): TE page 467; Answer Transparencies
____ Homework Quiz (9.3): TE page 469; Transparencies
____ Warm-Up: Transparencies

TEACHING OPTIONS

____ Notetaking Guide
____ Examples: 1–4, SE pages 470–472
____ Extra Examples: TE pages 471–472
____ Checkpoint Exercises: 1–11, SE pages 470–471
____ Concept Check: TE page 472
____ Guided Practice Exercises: 1–11, SE page 472

APPLY/HOMEWORK

Homework Assignment

____ Basic: SRH p. 773 Exs. 10–12, p. 784 Exs. 1–4;
     pp. 473–474 Exs. 12–21, 26–31, 33, 34, 36–39, 44–53
____ Average: pp. 473–474 Exs. 16–25, 28–37, 40–42, 44–53
____ Advanced: pp. 473–474 Exs. 18–25, 28–35, 38–44*, 47–53

Reteaching the Lesson

____ Practice: CRB pages 32–34 (Level A, Level B, Level C); Practice Workbook
____ Study Guide: CRB pages 35–36; Spanish Study Guide

Extending the Lesson

____ Challenge: SE page 474; CRB page 37

ASSESSMENT OPTIONS

____ Daily Quiz (9.4): TE page 474 or Transparencies
____ Standardized Test Practice: SE page 474
____ Quiz (9.1–9.4): SE page 475; Assessment Book page 107

Notes _____

LESSON
9.4

Teacher's Name _____ Class _____ Room _____ Date _____

Lesson Plan for Block Scheduling

Half-block lesson (See *Pacing and Assignment Guide*, TE page 450A)
For use with pages 470–474

GOAL Compare and order real numbers.

State/Local Objectives _____

_____

✓ **Check the items you wish to use for this lesson.**

STARTING OPTIONS
____ Homework Check (9.3): TE page 467; Answer Transparencies
____ Homework Quiz (9.3): TE page 469; Transparencies
____ Warm-Up: Transparencies

TEACHING OPTIONS
____ Notetaking Guide
____ Examples: 1–4, SE pages 470–472
____ Extra Examples: TE pages 471–472
____ Checkpoint Exercises: 1–11, SE pages 470–471
____ Concept Check: TE page 472
____ Guided Practice Exercises: 1–11, SE page 472

Chapter Pacing Guide	
Day	Lesson
1	9.1
2	9.2
3	9.3; **9.4**
4	9.5; 9.6 (begin)
5	9.6 (end); 9.7
6	9.8
7	Ch. 9 Review and Projects

APPLY/HOMEWORK
Homework Assignment
____ Block Schedule: pp. 473–474 Exs. 16–25, 28–37, 40–42, 44–53 (with 9.3)

Reteaching the Lesson
____ Practice: CRB pages 32–34 (Level A, Level B, Level C); Practice Workbook
____ Study Guide: CRB pages 35–36; Spanish Study Guide

Extending the Lesson
____ Challenge: SE page 474; CRB page 37

ASSESSMENT OPTIONS
____ Daily Quiz (9.4): TE page 474 or Transparencies
____ Standardized Test Practice: SE page 474
____ Quiz (9.1–9.4): SE page 475; Assessment Book page 107

Notes

_____

_____

_____

LESSON 9.4

Practice A

For use with pages 470–474

1. Are all real numbers rational?
2. A number n has a decimal form that does not terminate. What type of number is n if its decimal form repeats? What type of number is n if its decimal form doesn't repeat?

Tell whether the number is *rational* or *irrational*.

3. $\frac{3}{5}$
4. $\frac{1}{6}$
5. $\sqrt{5}$
6. $\sqrt{\frac{9}{4}}$
7. $-\sqrt{\frac{4}{7}}$
8. $0.\overline{3}$
9. $6.\overline{73}$
10. $-\frac{\sqrt{9}}{7}$

Complete the statement using <, >, or =.

11. $\frac{3}{2}$ __?__ $\sqrt{2.25}$
12. $-\sqrt{8}$ __?__ $-\frac{5}{2}$
13. $\sqrt{0.9}$ __?__ 0.9

Use a number line to order the numbers from least to greatest.

14. $3.4, \frac{\sqrt{16}}{2}, \sqrt{\frac{9}{2}}, \sqrt{10}$
15. $\sqrt{14}, \frac{19}{6}, \frac{\sqrt{24}}{2}, \sqrt{\frac{36}{3}}$
16. $9, \sqrt{75}, 7\sqrt{2}, 3\sqrt{8}$
17. $-11, -\sqrt{110}, -2\sqrt{37}, -\frac{21}{2}$

18. Name an irrational number between the rational numbers 9 and 10.

19. The area of a square is $\frac{9}{16}$ square meter. Is the length of one side of the square a *rational* or *irrational* number of meters? Explain.

20. You made a cardboard tree that is 48 inches tall to use as a stage prop for a play. You need to make a diagonal brace that connects a 28-inch long horizontal brace to the top of the tree. Find the length of the diagonal brace to the nearest tenth of an inch.

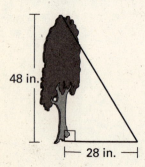

LESSON 9.4 Practice B
For use with pages 470–474

Tell whether the number is *rational* or *irrational*.

1. $\dfrac{1}{7}$
2. $\sqrt{\dfrac{1}{7}}$
3. $1.\overline{12}$
4. $-\sqrt{17}$
5. $-\sqrt{\dfrac{8}{2}}$
6. $\sqrt{\dfrac{21}{3}}$
7. $\dfrac{\sqrt{5}}{16}$
8. $\dfrac{\sqrt{16}}{5}$

Complete the statement using <, >, or =.

9. $\sqrt{\dfrac{1}{4}}$ ___?___ $\dfrac{1}{4}$
10. 4 ___?___ $\sqrt{\dfrac{32}{2}}$
11. $-\sqrt{8}$ ___?___ $-\dfrac{10}{3}$

In Exercises 12–15, use a number line to order the numbers from least to greatest.

12. $\dfrac{12}{11},\ \sqrt{1.1},\ \dfrac{\sqrt{10}}{3},\ \sqrt{\dfrac{10}{3}}$
13. $4.2,\ \sqrt{17},\ \dfrac{17}{4},\ \sqrt{\dfrac{81}{5}}$
14. $\sqrt{\dfrac{5}{9}},\ \dfrac{2}{3},\ \sqrt{\dfrac{7}{9}},\ 0.7514,\ 0.\overline{75}$
15. $\sqrt{31},\ 5.5,\ \dfrac{\sqrt{75}}{2},\ \dfrac{28}{5},\ \sqrt{\dfrac{55}{2}}$

16. One leg of a right triangle is 6 inches long. The other leg is 8 inches long. Is the length in inches of the hypotenuse a *rational* or *irrational* number?

17. From a balcony, you drop a penny 50 feet to the ground. The time t in seconds it takes the penny to hit the ground is approximated by $t = \sqrt{\dfrac{25}{8}}$. Does t represent a rational or an irrational number of seconds? Give the value of t to the nearest hundredth of a second.

18. You want to buy some frozen pizzas that have a diameter of 18 inches. You need to be able to fit them into your upright freezer that has a capacity of 4.5 cubic feet. The storage compartment of the freezer is 2 feet high. The width and depth of the compartment can be found using the expression $\sqrt{\dfrac{4.5}{2}}$. Will the pizzas lay flat on the shelves of your freezer? Explain.

19. Use a right triangle to graph $\sqrt{8}$ on a number line.

LESSON 9.4 Practice C
For use with pages 470–474

Tell whether the number is *rational* or *irrational*.

1. $\dfrac{\sqrt{1}}{\sqrt{4}}$
2. $\sqrt{\dfrac{4}{9}}$
3. $6.\overline{6}$
4. $-\sqrt{\dfrac{20}{3}}$
5. -9.61
6. $\dfrac{13}{2}$
7. $\sqrt{\dfrac{154}{7}}$
8. $\sqrt{\dfrac{294}{6}}$

Complete the statement using <, >, or =.

9. $\sqrt{\dfrac{84}{9}}\ \underline{?}\ 3$
10. $\sqrt{\dfrac{18}{64}}\ \underline{?}\ \dfrac{3\sqrt{2}}{8}$
11. $-\sqrt{\dfrac{7}{9}}\ \underline{?}\ -\dfrac{7}{9}$

In Exercises 12–15, use a number line to order the numbers from least to greatest.

12. $\dfrac{41}{6},\ \sqrt{44},\ 4\sqrt{3},\ 6.\overline{63}$
13. $3\sqrt{\dfrac{22}{7}},\ \sqrt{27},\ \dfrac{21}{4},\ 5.\overline{318}$
14. $-\sqrt{54},\ -2\sqrt{11},\ -\dfrac{4\sqrt{37}}{5},\ -\dfrac{81}{11}$
15. $\sqrt{\dfrac{15}{2}},\ \dfrac{\sqrt{15}}{2},\ 3\sqrt{0.9},\ \dfrac{57}{23},\ 2.\overline{47}$

16. You are making a square wooden game board with an area of 64 square inches. Is the length in *feet* of a side of the game board a rational number?

17. What has a greater perimeter, a square with an area of 217 square feet or a rectangle with an area of 200 square feet and a length of 25 feet? Explain.

18. You are standing on a cliff, 216 feet above the Pacific Ocean at Big Sur. You drop a rock into the surf. An equation you can use to find the time s in seconds for the rock to hit the surf is $s = \sqrt{\dfrac{216}{16}}$. Does it take the rock longer than $3\dfrac{1}{2}$ seconds to hit the surf? Explain.

19. An object is projected into the air. You calculate the length of time it takes the object to hit the ground to be $\dfrac{5\sqrt{5}}{2}$ seconds. Is there any way to write this exact number as a decimal? Explain.

20. Explain how you could use a triangle to graph $\sqrt{\dfrac{8}{9}}$.

LESSON 9.4

Name _____ Date _____

Study Guide
For use with pages 470–474

GOAL Compare and order real numbers.

> **VOCABULARY**
>
> An **irrational number** is a number that cannot be written as a quotient of two integers. The decimal form of an irrational number neither terminates nor repeats.
>
> The **real numbers** consist of all rational and irrational numbers.

EXAMPLE 1 Classifying Real Numbers

Number	Decimal Form	Decimal Type	Type
a. $2\frac{3}{4}$	$2\frac{3}{4} = 2.75$	Terminating	Rational
b. $-\frac{6}{11}$	$-\frac{6}{11} = -0.545454\ldots = -0.\overline{54}$	Repeating	Rational
c. $\sqrt{2}$	$\sqrt{2} = 1.41421356\ldots$	Nonterminating, nonrepeating	Irrational

Exercises for Example 1

Tell whether the number is *rational* or *irrational*.

1. $\frac{4}{9}$
2. -8
3. $-\sqrt{81}$
4. $\sqrt{11}$

EXAMPLE 2 Comparing Real Numbers

Copy and complete $\frac{14}{5}$ __?__ $\sqrt{5}$ using <, >, or =.

Solution

Graph $\frac{14}{5}$ and $\sqrt{5}$ on a number line.

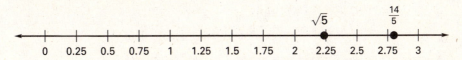

$\sqrt{5}$ is to the left of $\frac{14}{5}$.

Answer: $\frac{14}{5} > \sqrt{5}$

LESSON 9.4 Continued

Study Guide
For use with pages 470–474

EXAMPLE 3 Ordering Real Numbers

Use a number line to order the numbers $-\sqrt{7}, \frac{17}{5}, -3.1,$ and $\sqrt{6}$ from least to greatest.

Solution

Graph the numbers on a number line and read them from left to right.

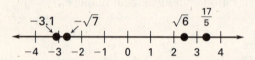

Answer: From least to greatest, the numbers are $-3.1, -\sqrt{7}, \sqrt{6},$ and $\frac{17}{5}$.

Exercises for Examples 2 and 3

Copy and complete the statement using <, >, or =.

5. $\sqrt{11}$ __?__ $\frac{11}{3}$

6. -9.3 __?__ $-\sqrt{82}$

7. $-\frac{39}{9}$ __?__ $-2\sqrt{5}$

8. $3\sqrt{2}$ __?__ $\sqrt{17}$

Use a number line to order the numbers from least to greatest.

9. $5\sqrt{2}, 7.1, \frac{32}{5}, \sqrt{21}$

10. $-2\sqrt{2}, -2.3, -\frac{13}{5}, -\sqrt{7}$

EXAMPLE 4 Using Irrational Numbers

A car in an accident leaves skid marks of 208 feet. Evaluate the expression $3\sqrt{3\ell}$, where ℓ is the length of the skid marks in feet, to find the car's speed in miles per hour at the time the brakes were applied.

Solution

$3\sqrt{3\ell} = 3\sqrt{3 \cdot 208}$ Substitute 208 for ℓ.

$= 3\sqrt{624}$ Multiply.

≈ 75 Approximate using a calculator.

Answer: The car's speed was about 75 miles per hour.

Exercises for Example 4

Using the information in Example 4 and the given length of the skid marks, find the speed of the car at the time the brakes were applied. Round your answer to the nearest mile per hour.

11. $\ell = 20$ ft

12. $\ell = 200$ ft

13. $\ell = 76$ ft

14. $\ell = 156$ ft

LESSON 9.4 Challenge Practice
For use with pages 470–474

Use a number line to order the numbers from least to greatest.

1. $4.7, \sqrt{17}, \frac{48}{11}, 4\frac{3}{5}, 4.15, 3\sqrt{2}$

2. $-3.5, -2\sqrt{2}, -\sqrt{7}, -\frac{187}{50}, -3\frac{4}{5}, -3.14$

3. $\frac{183}{20}, \sqrt{83}, 9\frac{1}{8}, 9.27, 4\sqrt{5}, 8.98$

4. $-0.87, -\sqrt{\frac{1}{2}}, -\frac{3}{4}, -\sqrt{0.25}, -0.71, -\frac{7}{16}$

Find an irrational number that is between the given rational numbers.

5. $\frac{1}{3}$ and $\frac{1}{2}$

6. $5\frac{1}{2}$ and $5\frac{5}{8}$

7. -4.25 and -4.2

8. Is the quotient of two irrational numbers *always*, *sometimes*, or *never* irrational? Give examples to support your answer.

LESSON
9.5

Teacher's Name _____ Class _____ Room _____ Date _____

Lesson Plan

1-day lesson (See *Pacing and Assignment Guide*, TE page 450A)
For use with pages 476–481

GOAL   Use the distance, midpoint, and slope formulas.

State/Local Objectives _____
_____

✓ **Check the items you wish to use for this lesson.**

STARTING OPTIONS

__ Homework Check (9.4): TE page 473; Answer Transparencies
__ Homework Quiz (9.4): TE page 474; Transparencies
__ Warm-Up: Transparencies

TEACHING OPTIONS

__ Notetaking Guide
__ Activity Master: CRB page 40
__ Examples: 1–4, SE pages 477–478
__ Extra Examples: TE pages 477–478
__ Checkpoint Exercises: 1–6, SE pages 477–478
__ Concept Check: TE page 478
__ Guided Practice Exercises: 1–6, SE page 479

APPLY/HOMEWORK

Homework Assignment

__ Basic: EP p. 830 Exs. 9–12, 32–37; pp. 479–481 Exs. 7–14, 19–24, 27–38, 40, 43–46, 55–62
__ Average: pp. 479–481 Exs. 9–18, 21–26, 30–35, 39–44, 47–53, 55–63
__ Advanced: pp. 479–481 Exs. 9–14, 21–29, 36–44, 47–56*, 60–63

Reteaching the Lesson

__ Practice: CRB pages 41–43 (Level A, Level B, Level C); Practice Workbook
__ Study Guide: CRB pages 44–45; Spanish Study Guide

Extending the Lesson

__ Real-World Problem Solving: CRB page 46
__ Challenge: SE page 481; CRB page 47

ASSESSMENT OPTIONS

__ Daily Quiz (9.5): TE page 481 or Transparencies
__ Standardized Test Practice: SE page 481

Notes _____
_____
_____

LESSON 9.5

Lesson Plan for Block Scheduling

Half-block lesson (See *Pacing and Assignment Guide*, TE page 450A)
For use with pages 476–481

Teacher's Name _____ Class _____ Room _____ Date _____

GOAL Use the distance, midpoint, and slope formulas.

State/Local Objectives _____
_____

✓ **Check the items you wish to use for this lesson.**

STARTING OPTIONS
____ Homework Check (9.4): TE page 473; Answer Transparencies
____ Homework Quiz (9.4): TE page 474; Transparencies
____ Warm-Up: Transparencies

TEACHING OPTIONS
____ Notetaking Guide
____ Activity Master: CRB page 40
____ Examples: 1–4, SE pages 477–478
____ Extra Examples: TE pages 477–478
____ Checkpoint Exercises: 1–6, SE pages 477–478
____ Concept Check: TE page 478
____ Guided Practice Exercises: 1–6, SE page 479

Chapter Pacing Guide	
Day	Lesson
1	9.1
2	9.2
3	9.3; 9.4
4	**9.5**; 9.6 (begin)
5	9.6 (end); 9.7
6	9.8
7	Ch. 9 Review and Projects

APPLY/HOMEWORK

Homework Assignment
____ Block Schedule: pp. 479–481 Exs. 9–18, 21–26, 30–35, 39–44, 47–53, 55–63 (with 9.6)

Reteaching the Lesson
____ Practice: CRB pages 41–43 (Level A, Level B, Level C); Practice Workbook
____ Study Guide: CRB pages 44–45; Spanish Study Guide

Extending the Lesson
____ Real-World Problem Solving: CRB page 46
____ Challenge: SE page 481; CRB page 47

ASSESSMENT OPTIONS
____ Daily Quiz (9.5): TE page 481 or Transparencies
____ Standardized Test Practice: SE page 481

Notes _____
_____
_____

LESSON 9.5 Activity Master

For use before Lesson 9.5

Goal
Use the Pythagorean theorem to find the distance between two points.

Materials
- pencil and paper
- graph paper

Finding the Distance Between Two Points

In this activity, you will discover a distance formula for the distance between two points in a coordinate plane.

INVESTIGATE Find the distance between points $A(1, 4)$ and $B(5, 1)$.

1. Draw a coordinate plane on a piece of graph paper and plot points $A(1, 4)$ and $B(5, 1)$.

2. Then plot point $C(1, 1)$. Connect the three points to form a right triangle, as shown.

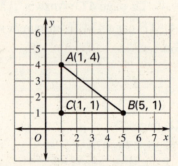

3. Points A and C lie on a vertical line. So, the distance between points A and C is the absolute value of the difference of their y-coordinates. Points B and C lie on a horizontal line. So, the distance between points B and C is the absolute value of the difference of their x-coordinates.

   $AC = |4 - 1| = 3$ and $BC = |5 - 1| = 4$

4. Use the Pythagorean theorem to find the distance between points A and B.

   $(AB)^2 = (AC)^2 + (BC)^2$
   $AB = \sqrt{(AC)^2 + (BC)^2}$
   $= \sqrt{3^2 + 4^2} = 5$

Answer: The distance between points $A(1, 4)$ and $B(5, 1)$ is 5 units.

DRAW CONCLUSIONS

Find the distance between the points. Write your answer in simplest form.

1. $A(2, 2)$, $B(6, 6)$
2. $A(1, 2)$, $B(2, 5)$
3. $A(3, 7)$, $B(15, 2)$
4. $A(0, 0)$, $B(3, -6)$

5. Suppose the coordinates for point A are (x_1, y_1) and the coordinates for point B are (x_2, y_2). What coordinates would you use for point C to form a right triangle, as in the activity?

6. Repeat steps 3 and 4 in the activity using the coordinates described in Exercise 5. Use your result to write a formula to find the distance between any two points in a plane.

Lesson 9.5 Practice A
For use with pages 476–481

Find the distance between the points. Write your answer in simplest form.

1. A and B
2. B and C
3. C and D
4. D and A

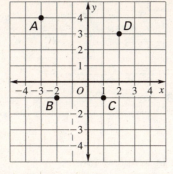

Find the distance between the points. Write your answer in simplest form.

5. (9, 5), (3, 1)
6. (1, 5), (−4, −3)
7. (−8, −6), (−5, 0)
8. (−2, −7), (−3, −1)
9. (−7, 4), (9, 1)
10. (−6, 5), (4, −3)

Find the midpoint of the segment.

11. $\overline{EF}$
12. $\overline{FG}$

Find the slope of the line through the given points.

13. D and E
14. G and H

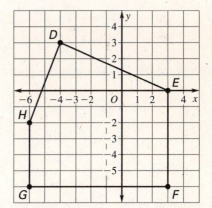

Find the midpoint of the segment with the given endpoints.

15. (−9, 1), (3, −2)
16. (7, −5), (−1, 3)
17. (4, 8), (6, 2)
18. (11, −10), (−12, 8)
19. (−8, −3), (−7, −4)
20. (−5, 5), (9, −6)

Find the slope of the line through the given points.

21. (−4, 7), (−2, 5)
22. (9, −3), (1, 0)
23. (−6, −4), (−3, −2)
24. (11, 10), (12, 4)
25. (−5, 7), (−2, −1)
26. (−10, 3), (−11, 7)

27. You are doing errands and plan to stop at the grocery store, the flower shop, and the bakery. A grid is superimposed on the map of your town.

   a. How far from the bakery is the grocery store?

   b. How far from the flower shop is the grocery store?

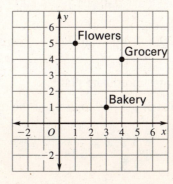

LESSON 9.5 Practice B

For use with pages 476–481

Find the distance between the points. Write your answer in simplest form.

1. A and B
2. B and C
3. C and D
4. D and A

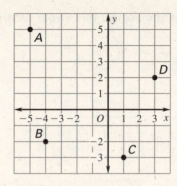

Find the distance between the points. Write your answer in simplest form.

5. $(-12, -8), (3, 11)$
6. $(0, 7), (-2, -1)$
7. $(1.5, -2), (-4, 6)$
8. $(4.9, -1), (7.4, -6)$
9. $(5.5, -3), (2.5, -7)$
10. $(4, -3.5), (-2, 8.5)$

Find the midpoint of the segment.

11. $\overline{GH}$
12. $\overline{HE}$
13. $\overline{EF}$
14. $\overline{FG}$

Find the slope of the line through the given points.

15. P and Q
16. Q and R
17. R and S
18. P and S

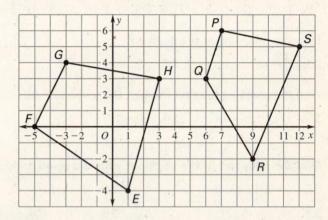

Find the midpoint of the segment with the given endpoints.

19. $(-9, -8), (1, 2)$
20. $(4, -5), (-2, 7)$
21. $(1.6, 0), (5.4, -3)$
22. $(-3.2, -1.2), (7.8, 6.8)$
23. $(-12, 5.4), (14, 3.6)$
24. $\left(15, -3\tfrac{1}{2}\right), (13, -1)$

Find the slope of the line through the given points.

25. $(8, -12), (10, -3)$
26. $(-9, 11), (7, -5)$
27. $(3.2, 7), (1.6, -9)$
28. $(6, -4.5), (1, -3.7)$
29. $\left(4\tfrac{3}{4}, -6\right), \left(1\tfrac{1}{4}, 5\right)$
30. $\left(12, \tfrac{9}{10}\right), \left(-11, \tfrac{3}{10}\right)$

31. The points $P(-4, 6)$, $Q(-4, -3)$, and $R(8, -3)$ are the vertices of a right triangle in a coordinate plane.

   a. Draw the triangle in a coordinate plane.

   b. Find the coordinates of the midpoint M of the hypotenuse of $\triangle PQR$.

LESSON 9.5 Practice C

For use with pages 476–481

Find the distance between the points. Write your answer in simplest form.

1. A and B
2. B and C
3. C and D
4. D and A

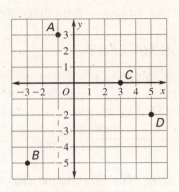

Find the distance between the points. Write your answer in simplest form.

5. (9.1, 4.2), (3.5, 7.4)
6. (6.2, 1.3), (2.4, 2.5)
7. (−4.5, −8.1), (−3.3, −1.9)
8. (−1.7, −2.6), (5.3, 4.6)
9. (4.1, 5.7), (−9.5, −6.3)
10. (−6.2, 7.8), (1.6, −2.4)

Find the midpoint of the segment.

11. $\overline{DE}$
12. $\overline{BH}$
13. $\overline{CB}$
14. $\overline{FG}$

Find the slope of the line through the given points.

15. C and D
16. E and F

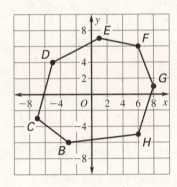

Find the midpoint of the segment with the given endpoints.

17. (2.5, −1.2), (−3.7, 4.8)
18. (−7.2, 12.1), (−9.4, 13.3)
19. (8.1, −4.7), (−3.5, −6.1)
20. (9.2, 12.1), (5.8, 14.3)
21. (20.1, 19.2), (15.3, 16.4)
22. (−13.6, 17.5), (−14.2, 11.5)

Find the slope of the line through the given points.

23. (−11.7, −12.5), (−19, −3)
24. (18, −4), (13.4, −15.1)
25. $\left(6\frac{3}{8}, -2\right), \left(-1\frac{5}{8}, 7\right)$

26. A bird watcher uses a grid to photograph and record the locations of birds feeding on the ground. The grid is made of net with strands that are 1 foot apart. In the grid shown, each point represents the location of a bird. How far apart are the birds at points A and F?

27. Tell which segment is longer. $\overline{AB}$ has endpoints A(7, 7) and B(2, −2). $\overline{PQ}$ has endpoints P(−6, 4) and Q(0, −4).

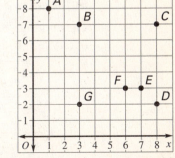

For $\overline{AB}$ with midpoint M, determine the coordinates of point B.

28. $A(-5, 6); M = \left(-1\frac{1}{2}, 1\right)$
29. $A(-3, -4); M = (-1, 0)$

30. Plot the points G(−6, −1), H(1, 3), and J(4, −1) in a coordinate plane and draw triangle GHJ. Determine whether the triangle is a right triangle.

LESSON 9.5

Name _____ **Date** _____

Study Guide
For use with pages 476–481

GOAL Use the distance, midpoint, and slope formulas.

VOCABULARY

The **midpoint** of a segment is the point on the segment that is equally distant from the endpoints. The coordinates of the midpoint of a segment are the average of the endpoints' x-coordinates and the average of the endpoints' y-coordinates.

EXAMPLE 1 Finding the Distance Between Two Points

Find the distance between the points $A(-3, 7)$ and $B(-8, -5)$.

$d = \sqrt{(x_2 - x_1)^2 + (y_2 - y_1)^2}$ Distance formula

$ = \sqrt{[-8 - (-3)]^2 + (-5 - 7)^2}$ Substitute -8 for x_2, -3 for x_1, -5 for y_2, and 7 for y_1.

$ = \sqrt{(-5)^2 + (-12)^2}$ Subtract.

$ = \sqrt{25 + 144}$ Evaluate powers.

$ = \sqrt{169}$ Add.

$ = 13$ Simplify.

Answer: The distance between the points $A(-3, 7)$ and $B(-8, -5)$ is 13 units.

Exercises for Example 1

Find the distance between the points.

1. $(1, 9), (8, 2)$ **2.** $(-5, -6), (6, -5)$ **3.** $(4, -4), (-8, 10)$

EXAMPLE 2 Using the Distance Formula

Use the city map to find the distance between the mall and the library. Each unit represents 1 mile.

Solution

The coordinates of the mall are $(7, 7)$. The coordinates of the library are $(6, 1)$.

$d = \sqrt{(x_2 - x_1)^2 + (y_2 - y_1)^2}$ Distance formula

$ = \sqrt{(6 - 7)^2 + (1 - 7)^2}$ Substitute 6 for x_2, 7 for x_1, 1 for y_2, and 7 for y_1.

$ = \sqrt{(-1)^2 + (-6)^2}$ Subtract.

$ = \sqrt{1 + 36}$ Evaluate powers.

$ = \sqrt{37}$ Add.

Answer: The distance between the mall and the library is $\sqrt{37}$, or about 6, miles.

LESSON 9.5 Continued

Name _____ Date _____

Study Guide
For use with pages 476–481

Exercises for Example 2

Use the information in Example 2 to find the distance between the locations.

4. gas station, park
5. park, library
6. mall, gas station

EXAMPLE 3 Finding a Midpoint

Find the midpoint M of the segment with endpoints $(-5, 9)$ and $(4, -3)$.

Solution

$M = \left(\dfrac{x_1 + x_2}{2}, \dfrac{y_1 + y_2}{2}\right)$ Midpoint formula

$= \left(\dfrac{-5 + 4}{2}, \dfrac{9 + (-3)}{2}\right)$ Substitute -5 for x_1, 4 for x_2, 9 for y_1, and -3 for y_2.

$= \left(-\dfrac{1}{2}, 3\right)$ Simplify.

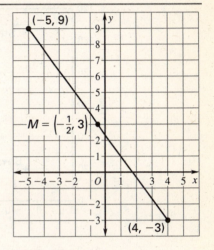

Exercises for Example 3

Find the midpoint M of the segment with the given endpoints.

7. $(10, 6), (2, 3)$
8. $(-5, 11), (7, -2)$
9. $(-11, -8), (-1, -7)$

EXAMPLE 4 Finding Slope

Find the slope of the line through $(-7, -2)$ and $(-3, 5)$.

Solution

$\text{slope} = \dfrac{y_2 - y_1}{x_2 - x_1}$ Slope formula

$= \dfrac{5 - (-2)}{-3 - (-7)}$ Substitute 5 for y_2, -2 for y_1, -3 for x_2, and -7 for x_1.

$= \dfrac{7}{4}$ Simplify.

Exercises for Example 4

Find the slope of the line through the given points.

10. $(8, 1), (10, -1)$
11. $(-6, -4), (3, -1)$
12. $(-7, 9), (-13, 14)$

LESSON 9.5

Real-World Problem Solving

For use with pages 476–481

Shipwreck

In Exercises 1–4, use the following information.

An underwater archaeology crew is mapping out a shipwreck on the ocean floor by laying a grid over the site. The ship is mostly submerged beneath the ocean, but the crew has labeled a few identifiable features, as shown below.

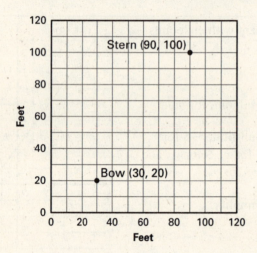

1. In the figure above, the front (bow) and rear (stern) of the ship are labeled. Use this information to find the length of the ship.

2. According to records, the mast of the ship is located halfway between the bow and the stern. Determine the coordinates where the mast should be located.

3. The crew finds the anchor of the ship located at the coordinates (120, 60). How far from the stern is the anchor located? How far from the bow is the anchor located? Round your answers to the nearest foot.

4. According to records, a storage compartment should be located $\frac{2}{5}$ of the ship's length from the bow. How many feet from the bow is the storage compartment?

LESSON 9.5
Challenge Practice
For use with pages 476–481

Plot the points A, B, and C in a coordinate plane and draw triangle ABC. Determine whether the triangle is a right triangle.

1. $A(2, 3)$, $B(6, 0)$, $C(8, 2)$

2. $A(-3, 3)$, $B(2, 2)$, $C(1, -2)$

Plot the points K, L, M, and N in a coordinate plane and draw rectangle KLMN. Then find the area and perimeter of the rectangle. If necessary, give your answers as radicals in simplest form.

3. $K(-2, 4)$, $L(4, -4)$, $M(12, 2)$, $N(6, 10)$

4. $K(1, 3)$, $L(5, 6)$, $M(6, 4)$, $N(2, 1)$

Let $\overline{AB}$ with endpoints A and B have midpoint M. Find the value of a.

5. $A(9, a + 3)$, $B(2, a - 1)$, $M\left(\dfrac{11}{2}, 6\right)$

6. $A(4a, 7)$, $B(8, -5)$, $M(a - 2, 1)$

7. $A(5, a + 1)$, $B(-11, -15)$, $M(-3, -3a)$

8. Divide the line segment with endpoints $A(0, 0)$ and $B(12, 8)$ into four segments of equal length. Give the endpoints of each segment.

9. The slope of the line through points $C(-3, -7)$ and $D(3, a^2 + 1)$ is $\dfrac{1}{2}a^2$. Find the value of a.

LESSON 9.6

Teacher's Name _____ Class _____ Room _____ Date _____

Lesson Plan

2-day lesson (See *Pacing and Assignment Guide*, TE page 450A)
For use with pages 482–487

GOAL Use special right triangles to solve problems.

State/Local Objectives _____

_____

✓ **Check the items you wish to use for this lesson.**

STARTING OPTIONS

____ Homework Check (9.5): TE page 479; Answer Transparencies
____ Homework Quiz (9.5): TE page 481; Transparencies
____ Warm-Up: Transparencies

TEACHING OPTIONS

____ Notetaking Guide
____ Concept Activity: SE page 482
____ Examples: Day 1: 1–2, SE pages 483–484; Day 2: 3, SE page 485
____ Extra Examples: TE pages 484–485
____ Checkpoint Exercises: Day 1: 1–3, SE page 484; Day 2: none
____ Concept Check: TE page 485
____ Guided Practice Exercises: Day 1: 1–5, SE page 485; Day 2: 6, SE page 485

APPLY/HOMEWORK

Homework Assignment

____ Basic: Day 1: pp. 486–487 Exs. 7–12, 22–26
Day 2: pp. 486–487 Exs. 13, 15–19, 27–31
____ Average: Day 1: pp. 486–487 Exs. 7–12, 22–27
Day 2: pp. 486–487 Exs. 13–20, 28–32
____ Advanced: Day 1: pp. 486–487 Exs. 7–12, 24–29
Day 2: pp. 486–487 Exs. 13–21*, 30–32

Reteaching the Lesson

____ Practice: CRB pages 50–52 (Level A, Level B, Level C); Practice Workbook
____ Study Guide: CRB pages 53–54; Spanish Study Guide

Extending the Lesson

____ Challenge: SE page 487; CRB page 55

ASSESSMENT OPTIONS

____ Daily Quiz (9.6): TE page 487 or Transparencies
____ Standardized Test Practice: SE page 487

Notes _____

LESSON 9.6

Teacher's Name _____ Class _____ Room _____ Date _____

Lesson Plan for Block Scheduling

1-block lesson (See *Pacing and Assignment Guide*, TE page 450A)
For use with pages 482–487

GOAL Use special right triangles to solve problems.

State/Local Objectives _____

_____

✓ **Check the items you wish to use for this lesson.**

STARTING OPTIONS

____ Homework Check (9.5): TE page 479; Answer Transparencies
____ Homework Quiz (9.5): TE page 481; Transparencies
____ Warm-Up: Transparencies

TEACHING OPTIONS

____ Notetaking Guide
____ Concept Activity: SE page 482
____ Examples: Day 4: 1–2, SE pages 483–484;
    Day 5: 3, SE page 485
____ Extra Examples: TE pages 484–485
____ Checkpoint Exercises: Day 4: 1–3, SE page 484; Day 5: none
____ Concept Check: TE page 485
____ Guided Practice Exercises: Day 4: 1–5, SE page 485; Day 5: 6, SE page 485

Chapter Pacing Guide

Day	Lesson
1	9.1
2	9.2
3	9.3; 9.4
4	9.5; **9.6 (begin)**
5	**9.6 (end)**; 9.7
6	9.8
7	Ch. 9 Review and Projects

APPLY/HOMEWORK

Homework Assignment

____ Block Schedule: Day 4: pp. 486–487 Exs. 7–12, 22–27 (with 9.5)
    Day 5: pp. 486–487 Exs. 13–20, 28–32 (with 9.7)

Reteaching the Lesson

____ Practice: CRB pages 50–52 (Level A, Level B, Level C); Practice Workbook
____ Study Guide: CRB pages 53–54; Spanish Study Guide

Extending the Lesson

____ Challenge: SE page 487; CRB page 55

ASSESSMENT OPTIONS

____ Daily Quiz (9.6): TE page 487 or Transparencies
____ Standardized Test Practice: SE page 487

Notes _____

LESSON 9.6 **Practice A**
For use with pages 482–487

Find the unknown length. Write your answer in simplest form.

1.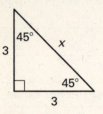
2.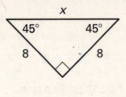
3.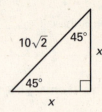
4.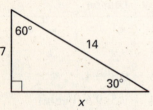
5.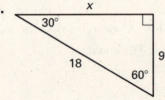
6.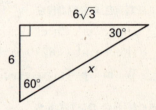

7. How is the length of the hypotenuse in a 45°-45°-90° triangle related to the length of a leg?

In Exercises 8 and 9, use the fact that the shorter leg of a 30°-60°-90° triangle is 5 inches to answer each question.

8. Find the length of the hypotenuse.

9. Find the length of the longer leg.

Find the unknown lengths. Write your answers in simplest form.

10.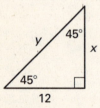
11.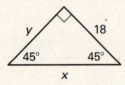
12.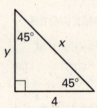
13.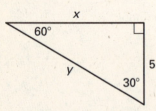
14.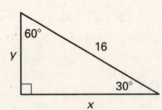
15.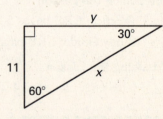

16. A staircase going down to the basement of a house is 15 feet long and makes a 45° angle with the basement floor. How many feet below the first floor is the basement floor? Round your answer to the nearest foot.

LESSON 9.6 Practice B

For use with pages 482–487

Find the unknown length. Write your answer in simplest form.

1.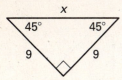
2.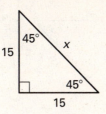
3.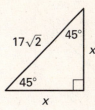
4.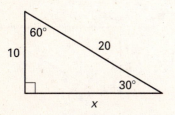
5.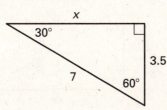
6.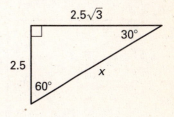

7. How is the length of the hypotenuse in a 30°-60°-90° triangle related to the length of the shorter leg?

Find the unknown lengths. Write your answers in simplest form.

8.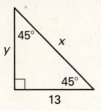
9.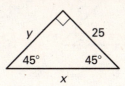
10.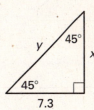
11.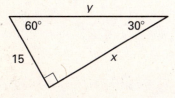
12.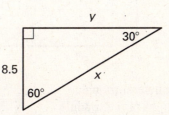
13.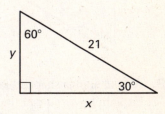

14. A 10-foot tipping trailer is being used to haul gravel. To unload the gravel, the front end of the trailer is raised. How high is the front end of a 10 foot tipping trailer when it is tipped by a 30° angle? by a 45° angle? Round your answers to the nearest foot.

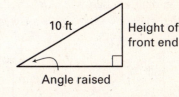

LESSON 9.6 Practice C

For use with pages 482–487

Find the unknown length. Write your answer in simplest form.

1.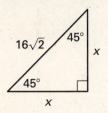
2.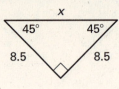
3.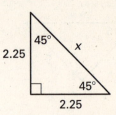

4.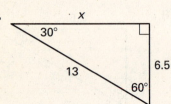
5.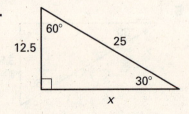
6.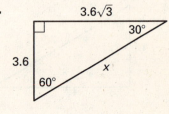

Find the unknown lengths. Write your answers in simplest form.

7.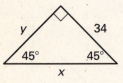
8.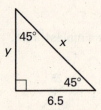
9.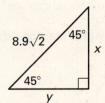

10.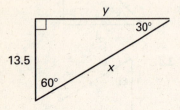
11.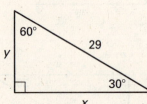
12.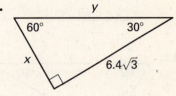

13. A road sign is shaped like an equilateral triangle. Explain how you can find the height of the triangle using special right triangles. Then estimate the height of the sign. Round your answer to the nearest inch.

    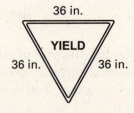

14. In the diagram below, $\triangle ABC \sim \triangle XYZ$. Find all the unknown side lengths of the triangles. Give exact answers.

    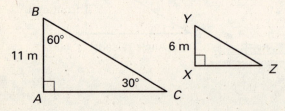

LESSON 9.6

Name _____ Date _____

Study Guide

For use with pages 482–487

GOAL Use special right triangles to solve problems.

EXAMPLE 1 Using a 45°-45°-90° Triangle

You are building corner shelves out of a square piece of wood. You cut along the diagonal of the square to produce two right triangles, each with a hypotenuse of $30\sqrt{2}$ centimeters. What is the side length x of each shelf?

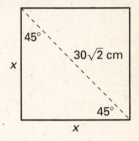

Solution

The diagonal divides the piece of wood into two 45°-45°-90° triangles. The diagonal is the hypotenuse of each triangle.

hypotenuse = leg • $\sqrt{2}$	Rule for 45°-45°-90° triangle
$30\sqrt{2} = x\sqrt{2}$	Substitute.
$\dfrac{30\sqrt{2}}{\sqrt{2}} = \dfrac{x\sqrt{2}}{\sqrt{2}}$	Divide each side by $\sqrt{2}$.
$30 = x$	Simplify.

Answer: The side length of each shelf is 30 centimeters.

EXAMPLE 2 Using a 30°-60°-90° Triangle

Find the length x of the shorter leg and the length y of the longer leg of the triangle.

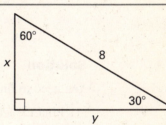

Solution

The triangle is a 30°-60°-90° triangle. The length of the hypotenuse is 8 units.

a. hypotenuse = 2 • shorter leg

$8 = 2x$ Substitute.

$\dfrac{8}{2} = \dfrac{2x}{2}$ Divide each side by 2.

$4 = x$ Simplify.

Answer: The length x of the shorter leg is 4 units.

b. longer leg = shorter leg • $\sqrt{3}$

$y = 4\sqrt{3}$ Substitute.

Answer: The length y of the longer leg is $4\sqrt{3}$ units.

LESSON 9.6 Continued

Study Guide

For use with pages 482–487

Exercises for Examples 1 and 2

Find the unknown lengths. Write your answers in simplest form.

1.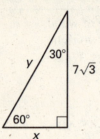
2.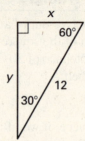
3.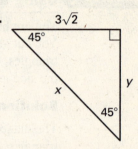

EXAMPLE 3 Using a Special Right Triangle

Your neighbors have a 30°-60°-90° triangular garden in the corner of their yard. The length of the longer leg of the triangle is $6\sqrt{3}$ feet. Find the lengths of the other sides of the triangle to the nearest foot.

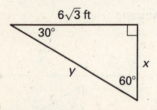

Solution

You need to find the length of the shorter leg first.

(1) Find the length x of the shorter leg.

longer leg = shorter leg · $\sqrt{3}$	Rule for 30°-60°-90° triangle
$6\sqrt{3} = x \cdot \sqrt{3}$	Substitute.
$6 = x$	Divide each side by $\sqrt{3}$.

(2) Find the length y of the hypotenuse.

hypotenuse = 2 · shorter leg	Rule for 30°-60°-90° triangle
$y = 2 \cdot 6$	Substitute.
$= 12$	Multiply.

Answer: The length of the shorter leg of the triangle is 6 feet. The length of the hypotenuse is 12 feet.

Exercise for Example 3

4. Find the lengths of the other two sides of the garden in Example 3 if the hypotenuse is 16 feet.

LESSON 9.6 Challenge Practice

For use with pages 482–487

Find the unknown lengths. Write your answers in simplest form.

1.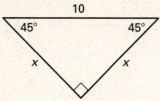

2.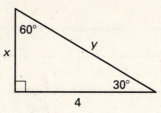

3.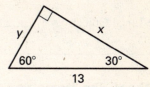

4.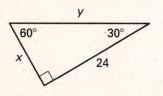

5.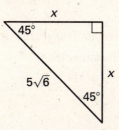

6.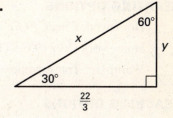

Find the value of x. Give your answer as a radical in simplest form.

7.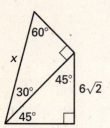

8.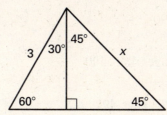

9.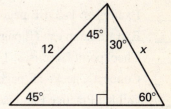

10. The coordinates of a 45°-45°-90° triangle are $A(2, 1)$, $B(-2, 5)$, and $C(6, 5)$. Find the lengths of the sides of the triangle. Explain how you found the lengths.

LESSON
9.7

Teacher's Name _____ Class _____ Room _____ Date _____

Lesson Plan
1-day lesson (See *Pacing and Assignment Guide*, TE page 450A)
For use with pages 488–493

GOAL   Use tangent to find side lengths of right triangles.

State/Local Objectives _____

_____

✓ **Check the items you wish to use for this lesson.**

STARTING OPTIONS
____ Homework Check (9.6): TE page 486; Answer Transparencies
____ Homework Quiz (9.6): TE page 487; Transparencies
____ Warm-Up: Transparencies

TEACHING OPTIONS
____ Notetaking Guide
____ Concept Activity: SE page 488
____ Examples: 1–3, SE pages 489–490
____ Extra Examples: TE page 490
____ Checkpoint Exercises: 1–2, SE pages 489–490
____ Concept Check: TE page 490
____ Guided Practice Exercises: 1–6, SE page 491

APPLY/HOMEWORK
Homework Assignment
____ Basic: pp. 491–493 Exs. 7–13, 18–21, 23–26, 28, 31–38
____ Average: pp. 491–493 Exs. 7–9, 14–29, 31–38
____ Advanced: pp. 491–493 Exs. 7–9, 14–19, 21–31*, 34–38

Reteaching the Lesson
____ Practice: CRB pages 58–60 (Level A, Level B, Level C); Practice Workbook
____ Study Guide: CRB pages 61–62; Spanish Study Guide

Extending the Lesson
____ Real-World Problem Solving: CRB page 63
____ Challenge: SE page 493; CRB page 64

ASSESSMENT OPTIONS
____ Daily Quiz (9.7): TE page 493 or Transparencies
____ Standardized Test Practice: SE page 493

Notes _____

_____

LESSON 9.7

Teacher's Name _____ Class _____ Room _____ Date _____

Lesson Plan for Block Scheduling
Half-block lesson (See *Pacing and Assignment Guide*, TE page 450A)
For use with pages 488–493

GOAL Use tangent to find side lengths of right triangles.

State/Local Objectives _____

_____

✓ **Check the items you wish to use for this lesson.**

Chapter Pacing Guide	
Day	Lesson
1	9.1
2	9.2
3	9.3; 9.4
4	9.5; 9.6 (begin)
5	9.6 (end); **9.7**
6	9.8
7	Ch. 9 Review and Projects

STARTING OPTIONS
____ Homework Check (9.6): TE page 486; Answer Transparencies
____ Homework Quiz (9.6): TE page 487; Transparencies
____ Warm-Up: Transparencies

TEACHING OPTIONS
____ Notetaking Guide
____ Concept Activity: SE page 488
____ Examples; 1–3, SE pages 489–490
____ Extra Examples: TE page 490
____ Checkpoint Exercises: 1–2, SE pages 489–490
____ Concept Check: TE page 490
____ Guided Practice Exercises: 1–6, SE page 491

APPLY/HOMEWORK
Homework Assignment
____ Block Schedule: pp. 491–493 Exs. 7–9, 14–29, 31–38 (with 9.6)

Reteaching the Lesson
____ Practice: CRB pages 58–60 (Level A, Level B, Level C); Practice Workbook
____ Study Guide: CRB pages 61–62; Spanish Study Guide

Extending the Lesson
____ Real-World Problem Solving: CRB page 63
____ Challenge: SE page 493; CRB page 64

ASSESSMENT OPTIONS
____ Daily Quiz (9.7): TE page 493 or Transparencies
____ Standardized Test Practice: SE page 493

Notes _____

_____

_____

LESSON 9.7 Practice A

For use with pages 488–493

Complete the statement.

1. In a right triangle, the __?__ of an acute angle is the ratio of the length of the side opposite the angle to the length of the side adjacent to the angle.

2. A __?__ ratio is a ratio of the lengths of two sides of a right triangle.

Find tan A. Write your answer as a fraction in simplest form.

3.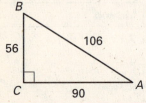
4.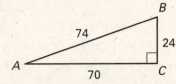
5.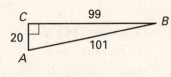

Use a calculator to approximate the tangent value to four decimal places.

6. tan 21°
7. tan 88°
8. tan 62°
9. tan 2°
10. tan 50°
11. tan 72°

Use the table of trigonometric ratios on page 823 to write the value of the tangent.

12. tan 40°
13. tan 25°
14. tan 35°

Find the value of x. Round to the nearest tenth.

15.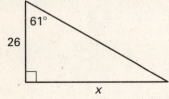
16.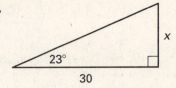
17.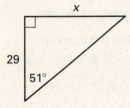

18. You are standing 100 feet from a tree. You estimate that the angle of elevation from your feet to the top of the tree is 50°. About how tall is the tree?

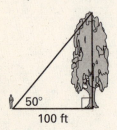

19. Suppose the tangent of an acute angle in a right triangle is less than 1. How does the side opposite the angle compare to the side adjacent to the angle?

LESSON 9.7 Practice B

For use with pages 488–493

Find the tangent of each acute angle. Write your answers as fractions in simplest form.

1.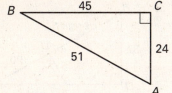
2.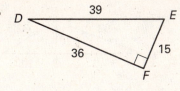
3.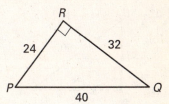

Use a calculator to approximate the tangent value to four decimal places.

4. tan 32°
5. tan 68°
6. tan 43°
7. tan 76°
8. tan 14°
9. tan 82°

Use the table of trigonometric ratios on page 823 to write the value of the tangent.

10. tan 22°
11. tan 56°
12. tan 39°

In Exercises 13–15, find the value of x. Round to the nearest tenth.

13.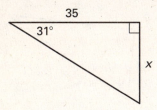
14.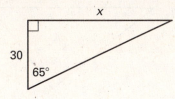
15.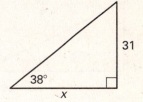

16. A hot air balloon climbs, at a 30° angle to the ground, to a height of 800 feet. To the nearest tenth of a foot, what ground distance has the balloon traveled to reach 800 feet?

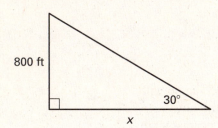

17. You are standing 80 feet from the base of a building. You estimate that the angle of elevation from your feet to the top of the building is about 70°. About how tall is the building?

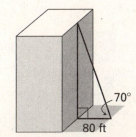

LESSON 9.7

Practice C

For use with pages 488–493

Find the tangent of each acute angle. Write your answers as fractions in simplest form.

1.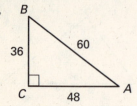
2.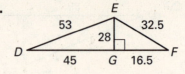
3.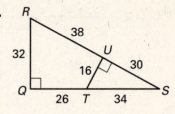

Use a calculator to approximate the tangent value to four decimal places.

4. tan 55.5°   5. tan 34.6°   6. tan 21.1°
7. tan 0.3°   8. tan 11.9°   9. tan 89.9°

Use the table of trigonometric ratios on page 823 to write the value of the tangent.

10. tan 42°   11. tan 18°   12. tan 37°

In Exercises 13–15, find the value of x. Round to the nearest tenth.

13.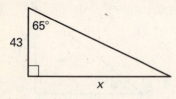
14.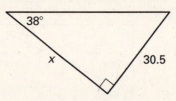
15.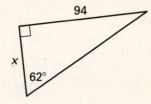

16. In the triangle shown below, find the value of x to the nearest unit. Then use the value of x to find the value of y to the nearest unit.

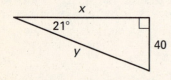

17. A loading ramp to the entrance of a building should make a 15° angle with the ground. If the bottom of the door is 6 feet above the ground, how far away from the entrance should the ramp begin? Round to the nearest tenth of a foot.

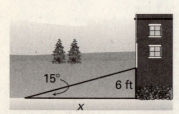

18. What is the largest value possible for the tangent of an acute angle of a triangle? Explain your reasoning.

LESSON 9.7

Name _____ Date _____

Study Guide
For use with pages 488–493

GOAL Use tangent to find side lengths of right triangles.

VOCABULARY

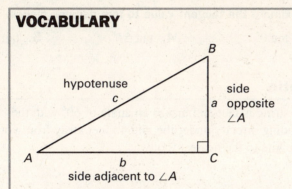

A **trigonometric ratio** is a ratio of the lengths of two sides of a right triangle. The **tangent** of an acute angle of a right triangle is the ratio of the length of the side opposite the angle to the length of the side adjacent to the angle.

$$\tan A = \frac{\text{side opposite } \angle A}{\text{side adjacent to } \angle A} = \frac{a}{b}$$

EXAMPLE 1 Finding a Tangent Ratio

For $\triangle ABC$, find the tangent of $\angle A$.

Solution

$$\tan A = \frac{\text{opposite}}{\text{adjacent}} = \frac{9}{12} = \frac{3}{4}$$

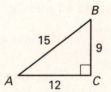

Exercise for Example 1

1. For $\triangle ABC$ in Example 1, find $\tan B$.

EXAMPLE 2 Using a Calculator

Use a calculator to approximate $\tan 80°$.

Keystrokes

2nd [TRIG]   = 80) =

Display

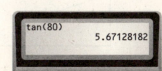

Answer: $\tan 80° \approx 5.6713$

LESSON 9.7 Continued

Study Guide
For use with pages 488–493

Exercises for Example 2

Use a calculator to approximate the tangent value to four decimal places.

2. tan 10°   3. tan 0°   4. tan 57°   5. tan 35°

EXAMPLE 3 Using a Tangent Ratio

You are flying a kite. The string is taut and makes an angle of 60° with the ground. Your friend is standing directly under the kite 40 feet away from you. What is the height h of the kite to the nearest foot?

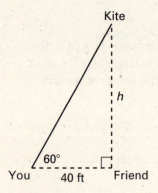

Solution

Use the tangent ratio. In the diagram, the length of the leg opposite the 60° angle is h. The length of the adjacent leg is 40 feet.

$\tan 60° = \dfrac{\text{opposite}}{\text{adjacent}}$ Definition of tangent ratio

$\tan 60° = \dfrac{h}{40}$ Substitute.

$1.7321 \approx \dfrac{h}{40}$ Use a calculator to approximate tan 60°.

$69.284 \approx h$ Multiply each side by 40.

Answer: The height of the kite is about 69 feet.

Exercises for Example 3

Find the value of x. Round to the nearest tenth.

6.

```
   |\
 x |  \
   |____\  18°
     45
```

7.

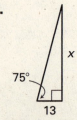

LESSON 9.7 Real-World Problem Solving

For use with pages 488–493

Creating a Scale Model

Your class is helping the local museum create a scale model of your town. The class is helping by recording the heights of different landmark buildings in town. To do this, you are measuring the angle of elevation from a point on the ground to the top of the building.

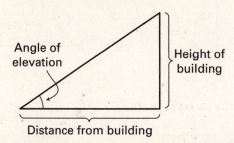

In Exercises 1 and 2, use the table that shows the measurements.

Building	Angle of Elevation	Distance from Building
Library	60°	48 feet
Elementary School	32°	40 feet
Middle School	52°	45 feet
High School	48°	60 feet
Town Hall	60°	20 feet
Bank	50°	25 feet
Supermarket	43°	30 feet

1. Find the heights of the buildings. Round your answers to the nearest foot.

2. The museum is planning on using a scale of 1 in. : 6 ft to make the scale model of the town. Find the heights of the buildings in the scale model. Round your answers to the nearest inch.

LESSON 9.7 Challenge Practice
For use with pages 488–493

Find the missing length. Then find tan A. If necessary, give your answers as radicals in simplest form.

1.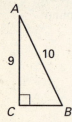

2.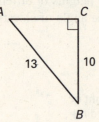

3.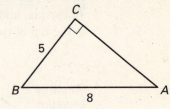

Find the area of the triangle. Round your answer to the nearest tenth.

4.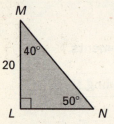

5.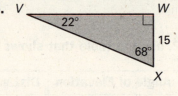

6.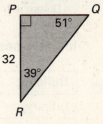

The figures are composed of rectangles and triangles. Find the area of the figure. Round your answer to the nearest tenth.

7.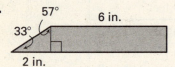

8.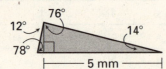

9.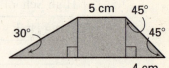

LESSON 9.8

Teacher's Name _____ Class _____ Room _____ Date _____

Lesson Plan

2-day lesson (See *Pacing and Assignment Guide*, TE page 450A)
For use with pages 494–499

GOAL Use sine and cosine to find triangle side lengths.

State/Local Objectives _____

_____

✓ **Check the items you wish to use for this lesson.**

STARTING OPTIONS

____ Homework Check (9.7): TE page 491; Answer Transparencies
____ Homework Quiz (9.7): TE page 493; Transparencies
____ Warm-Up: Transparencies

TEACHING OPTIONS

____ Notetaking Guide
____ Examples: Day 1: 1–2, SE pages 494–495; Day 2: 3–4, SE page 495
____ Extra Examples: TE page 495
____ Checkpoint Exercises: Day 1: 1, SE page 495; Day 2: none
____ Concept Check: TE page 496
____ Guided Practice Exercises: Day 1: 1–2, SE page 496; Day 2: 3–4, SE page 496
____ Technology Activity: SE page 499

APPLY/HOMEWORK

Homework Assignment

____ Basic: Day 1: pp. 497–498 Exs. 5–15, 30–35; Day 2: pp. 497–498 Exs. 16–22, 24–26, 36–39
____ Average: Day 1: pp. 497–498 Exs. 5–15, 30–36; Day 2: pp. 497–498 Exs. 16–28, 37–39
____ Advanced: Day 1: pp. 497–498 Exs. 5–15, 31–35; Day 2: pp. 497–498 Exs. 18–29*, 38, 39

Reteaching the Lesson

____ Practice: CRB pages 67–69 (Level A, Level B, Level C); Practice Workbook
____ Study Guide: CRB pages 70–71; Spanish Study Guide

Extending the Lesson

____ Challenge: SE page 498; CRB page 72

ASSESSMENT OPTIONS

____ Daily Quiz (9.8): TE page 498 or Transparencies
____ Standardized Test Practice: SE page 498
____ Quiz (9.5–9.8): Assessment Book page 108

Notes _____

_____

LESSON
9.8

Teacher's Name _____ Class _____ Room _____ Date _____

Lesson Plan for Block Scheduling
1-block lesson (See *Pacing and Assignment Guide*, TE page 450A)
For use with pages 494–499

GOAL Use sine and cosine to find triangle side lengths.

State/Local Objectives _____
_____

✓ **Check the items you wish to use for this lesson.**

Chapter Pacing Guide	
Day	Lesson
1	9.1
2	9.2
3	9.3; 9.4
4	9.5; 9.6 (begin)
5	9.6 (end); 9.7
6	**9.8**
7	Ch. 9 Review and Projects

STARTING OPTIONS
____ Homework Check (9.7): TE page 491; Answer Transparencies
____ Homework Quiz (9.7): TE page 493; Transparencies
____ Warm-Up: Transparencies

TEACHING OPTIONS
____ Notetaking Guide
____ Examples: 1–4, SE pages 494–495
____ Extra Examples: TE page 495
____ Checkpoint Exercises: 1, SE page 495
____ Concept Check: TE page 496
____ Guided Practice Exercises: 1–4, SE page 496
____ Technology Activity: SE page 499

APPLY/HOMEWORK

Homework Assignment
____ Block Schedule: pp. 497–498 Exs. 5–28, 30–39

Reteaching the Lesson
____ Practice: CRB pages 67–69 (Level A, Level B, Level C); Practice Workbook
____ Study Guide: CRB pages 70–71; Spanish Study Guide

Extending the Lesson
____ Challenge: SE page 498; CRB page 72

ASSESSMENT OPTIONS
____ Daily Quiz (9.8): TE page 498 or Transparencies
____ Standardized Test Practice: SE page 498
____ Quiz (9.5–9.8): Assessment Book page 108

Notes _____
_____
_____

LESSON 9.8 Practice A

For use with pages 494–499

Find the sine and cosine of each acute angle. Write your answers in simplest form.

1.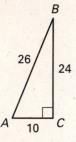
2.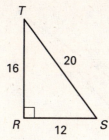
3.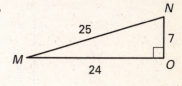

Approximate the sine or cosine value to four decimal places.

4. cos 60°
5. sin 19°
6. cos 24°
7. cos 73°
8. sin 53°
9. sin 85°
10. cos 13°
11. sin 67°
12. cos 3°

Find the value of x to the nearest tenth.

13.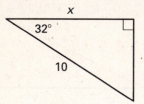
14.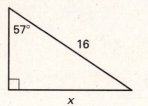
15.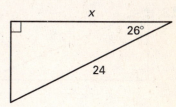
16.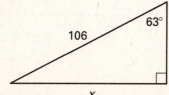
17.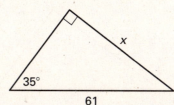
18.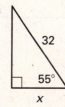

19. A diagram of a ski jump is shown below. To the nearest meter, what is the height h of the ski jump?

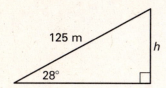

LESSON 9.8 Practice B
For use with pages 494–499

Find the sine and cosine of each acute angle. Write your answers in simplest form.

1.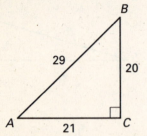
2.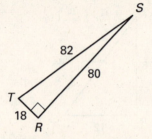
3.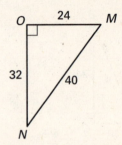

Approximate the sine or cosine value to four decimal places.

4. sin 45°
5. cos 17°
6. sin 25°
7. cos 37°
8. sin 35°
9. cos 58°
10. sin 8°
11. cos 62°
12. sin 44°

Find the value of x to the nearest tenth.

13.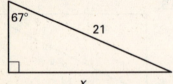
14.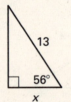
15.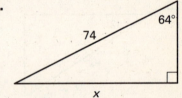
16.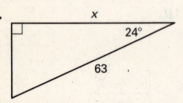
17.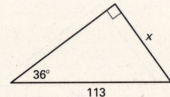
18.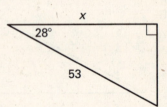

19. A water slide is 70.3 meters long and makes an angle of 19° with the ground. To the nearest tenth of a meter, what is the height h of the water slide?

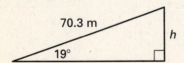

LESSON 9.8 Practice C

For use with pages 494–499

Find the sine and cosine of each acute angle. Write your answers in simplest form.

1.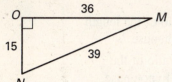
2.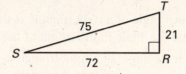
3.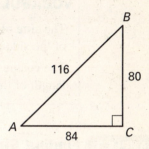

Approximate the sine or cosine value to four decimal places.

4. cos 35°
5. sin 6°
6. cos 52°
7. sin 10°
8. sin 62°
9. cos 49°
10. sin 21°
11. cos 26°
12. sin 41°

Find the value of x to the nearest tenth.

13.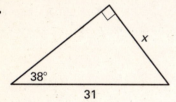
14.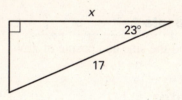
15.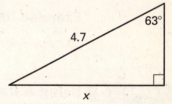
16.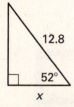
17.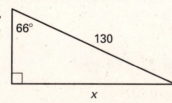
18.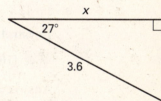

19. A skateboard ramp is 2.3 meters long and makes an angle of 19.5° with the ground. Find the lengths of the legs of the triangle that supports the ramp. Round your answers to the nearest tenth of a meter.

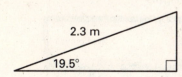

20. Can the values for the sine or cosine of one of the acute angles in a right triangle be greater than or equal to 1? Explain.

LESSON
9.8 Study Guide

Name _____ Date _____

For use with pages 494–499

GOAL Use sine and cosine to find triangle side lengths.

> **VOCABULARY**
>
> The **sine** of an acute angle of a right triangle is the ratio of the length of the side opposite the angle to the length of the hypotenuse.
>
> The **cosine** of an acute angle of a right triangle is the ratio of the length of the angle's adjacent side to the length of the hypotenuse.

EXAMPLE 1 Finding Sine and Cosine Ratios

For $\triangle MNP$, find the sine and cosine of $\angle M$.

Solution

$\sin M = \dfrac{\text{opposite}}{\text{hypotenuse}} = \dfrac{7}{25}$

$\cos M = \dfrac{\text{adjacent}}{\text{hypotenuse}} = \dfrac{24}{25}$

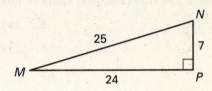

Exercise for Example 1

1. For $\triangle MNP$, find the sine and cosine of $\angle N$.

EXAMPLE 2 Using a Calculator

a. $\sin 12°$

Keystrokes

Display

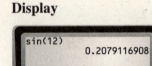

Answer: $\sin 12° \approx 0.2079$

b. $\cos 5°$

Keystrokes

Display

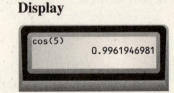

Answer: $\cos 5° \approx 0.9962$

Exercises for Example 2

Use a calculator to approximate the sine or cosine value to four decimal places.

2. $\sin 17°$ 3. $\sin 67°$ 4. $\cos 35°$ 5. $\cos 85°$

LESSON 9.8 Continued

Study Guide
For use with pages 494–499

EXAMPLE 3 Using a Cosine Ratio

In $\triangle GHJ$ shown, $\overline{GJ}$ is adjacent to $\angle G$. You know the length of the hypotenuse. To find the value of x, use cos G.

$\cos G° = \dfrac{\text{adjacent}}{\text{hypotenuse}}$ Definition of cosine ratio

$\cos 32° = \dfrac{x}{11}$ Substitute.

$0.8480 \approx \dfrac{x}{11}$ Use a calculator to approximate cos 32°.

$9.328 \approx x$ Multiply each side by 11.

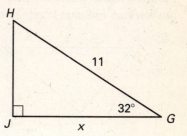

EXAMPLE 4 Using a Sine Ratio

In $\triangle ABC$ shown, $\overline{BC}$ is opposite $\angle A$. You know the length of the hypotenuse. To find the value of x, use sin A.

$\sin A° = \dfrac{\text{opposite}}{\text{hypotenuse}}$ Definition of sine ratio

$\sin 43° = \dfrac{x}{32}$ Substitute.

$0.6820 \approx \dfrac{x}{32}$ Use a calculator to approximate sin 43°.

$21.824 \approx x$ Multiply each side by 32.

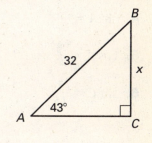

Exercises for Examples 3 and 4

Find the value of x to the nearest tenth.

6.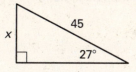

7.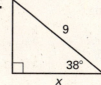

LESSON 9.8 Challenge Practice

For use with pages 494–499

Find the missing length. Then find sin A and cos A. If necessary, give your answers as radicals in simplest form.

1.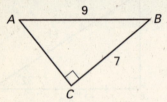

2.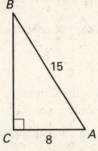

3.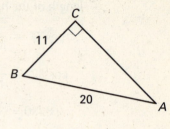

Find the possible lengths of the sides of right triangle ABC having the given sine or cosine. If necessary, give your answers as radicals in simplest form

4. $\sin A = \dfrac{4}{5}$

5. $\cos A = \dfrac{8}{13}$

6. $\sin C = \dfrac{7}{15}$

7. Suppose that angle A in a right triangle has a measure that is less than 45°. Which is greater, sin A or cos A? Use a triangle to help explain your reasoning.

8. Suppose that angle A in a right triangle has a measure that is greater than 45°, but is not 90°. Which is greater, sin A or cos A? Use a triangle to help explain your reasoning.

CHAPTER 9

Name _____ Date _____

Chapter Review Games and Activities

For use after Chapter 9

Word Scramble

Some of the vocabulary words from this chapter are listed below, but the order of the letters has been scrambled. Rearrange the letters in each scramble, using the empty boxes below the scrambled word. Then give an example of the word.

1. O N I R T M G O T R C I E   T R I A O

2. C L D A I A R   N E P X S S E I O R

3. G E L

4. L R E A   B M R E N U

5. N E O I S C

6. E R A S U Q   O T R O

7. G N H R P O Y A E A T   E M R H T O E

8. M N O D I I T P

9. T N N G T A E

10. E R T F C E P   E S U R A Q

11. I N E S

12. N E T P H U S Y O E

13. L T I N I R A O A R   M R N E B U

CHAPTER 9

Name _____ Date _____

Real-Life Project: Archeology
For use after Chapter 9

Objective Study the map of an archeological site.

Materials pencil, paper, graph paper, straightedge, colored pencils

Investigation *Getting Going* Archeologists can study the lives and cultures of people from the past by examining such evidence as the ruins of ancient cities, artwork, or hand tools. An archeological site is an area where evidence of past civilizations is found. Archeologists usually make maps of these sites to help with their description and to mark the locations of their findings.

A team of archeologists is studying a site that has been recently discovered. To organize their results, the team uses a coordinate plane to map the findings at the site.

Questions

1. The first evidence found at the site is a main chamber. The map is going to be centered around the main chamber that has coordinates $A(-0.5, 1)$, $B(0.5, 1)$, $C(1, -1)$, and $D(-1, -1)$. Use graph paper to draw a coordinate plane. Graph the coordinates of the chamber and draw $ABCD$.

2. Use the distance formula to find the perimeter of $ABCD$. Round your answer to the nearest tenth.

3. Four passages are discovered that lead out of the main chamber. The endpoints of each passage are given. For each passage, plot and connect the endpoints, and find its length. Then order the passages from shortest to longest. If necessary, round the length to the nearest tenth.

   a. Passage 1: $B(0.5, 1)$ and $E(7, 6)$

   b. Passage 2: $C(1, -1)$ and $F(4, -7)$

   c. Passage 3: $D(-1, -1)$ and $G(-8, -3)$

   d. Passage 4: $A(-0.5, 1)$ and $H(-2, 9)$

4. The archeologists discover wall drawings at the midpoint of each passage. Find and graph the midpoints of each passage to show where the wall drawings are located. Use a blue colored pencil to plot the midpoints.

5. Each passage branches off into many different rooms. The points below show where artifacts have been found. Use a red colored pencil and plot each point on the coordinate plane.

   a. $J(3.5, 5)$ b. $K(5, -4)$ c. $L(-4, -5)$ d. $M(-4.5, 4)$

6. What do you notice about the points where the artifacts were found?

Teacher's Notes for the Archeology Project

For use after Chapter 9

Project Goals
- Plot points in the coordinate plane.
- Find the distance between points.
- Find the midpoint of a segment.
- Evaluate square roots.

Managing the Project

Guiding Students' Work There is much information that will be graphed on the student's coordinate plane. Remind students to make their coordinate planes large enough so that everything will fit and to use colored pencils so that the coordinate plane is easier to understand.

Encourage students to show all of their work.

Rubric for Project

The following rubric can be used to assess student work.

4 The student draws a coordinate plane and correctly plots the points for the main chamber. The perimeter is correct. The lengths of each passage are correct and ordered correctly. All of the midpoints are correct. All of the calculations are correct and the student's work is shown. The coordinate plane is correctly labeled and neat. The student's work is organized and neat.

3 The student draws a coordinate plane and correctly plots the points for the main chamber. The perimeter is correct. The lengths of each passage are given and ordered but there are some minor errors. The midpoints are given but there are one or two minor errors. Most of the calculations are correct. The coordinate plane is labeled with some minor errors. The student's work is neat.

2 The student draws a coordinate plane and plots the points for the main chamber. The student has difficulty finding the perimeter. The lengths of each passage are given and ordered but there are some errors. The midpoints are given but there is more than one error. Some of the calculations are incorrect. The coordinate plane is labeled but there are some errors. The student's work is a little sloppy.

1 The student draws a coordinate plane and plots the points for the main chamber. The perimeter is incorrectly calculated. The lengths of each passage are incorrect. Some of the midpoints are incorrect. Most of the calculations are incorrect. The coordinate plane is incomplete and missing some labels. The student's work is incomplete or sloppy.

CHAPTER 9

Name _____ **Date** _____

Cooperative Project: Card Game

For use after Chapter 9

Objective Create and play a game that helps you practice ordering real numbers.

Materials cardboard or heavy paper, scissors, markers, pencils, paper

Investigation *Setting up the Game* To make the game cards, cut the cardboard into 60 rectangles that are the same size. On these cards, write the following real numbers. Write one real number on each card.

$\sqrt{13}$	$\dfrac{5}{\sqrt{16}}$	$-2\dfrac{1}{3}$	$-\dfrac{4}{9}$	$2\sqrt{4}$	$11\dfrac{2}{5}$
$-\dfrac{5}{3}$	$\sqrt{32}$	$-2\sqrt{9}$	$3\dfrac{3}{5}$	$\sqrt{\dfrac{144}{49}}$	$-\sqrt{48}$
$2\sqrt{25}$	$-\dfrac{\sqrt{49}}{8}$	$-\sqrt{104}$	$\dfrac{6}{11}$	$-\dfrac{25}{4}$	$\dfrac{\sqrt{81}}{12}$
$6\dfrac{2}{3}$	$\dfrac{4}{15}$	$-8\dfrac{5}{7}$	$-\sqrt{\dfrac{4}{9}}$	$\sqrt{3}$	$0.1\sqrt{36}$
$-\sqrt{18}$	$-\dfrac{3}{\sqrt{4}}$	$7\dfrac{3}{8}$	$\sqrt{120}$	$-1\dfrac{7}{8}$	$-\dfrac{13}{6}$
$1\dfrac{4}{9}$	$-\sqrt{11}$	$-\dfrac{2}{3}$	$-\dfrac{\sqrt{64}}{18}$	$3\dfrac{4}{5}$	$0.8\sqrt{9}$
$-\sqrt{5}$	$\dfrac{\sqrt{625}}{5}$	$\dfrac{11}{8}$	$-\sqrt{165}$	$\dfrac{8}{3}$	$-\sqrt{31}$
$\sqrt{\dfrac{100}{9}}$	$\dfrac{16}{3}$	$1\dfrac{3}{4}$	$\dfrac{\sqrt{25}}{3}$	$\sqrt{7}$	$-4\dfrac{1}{8}$
$-\dfrac{1}{9}$	$\sqrt{10}$	$\dfrac{15}{2}$	$\sqrt{156}$	$-3\sqrt{16}$	$\sqrt{\dfrac{16}{81}}$
$2\dfrac{1}{4}$	$-\dfrac{\sqrt{49}}{21}$	$-\dfrac{17}{2}$	$-\sqrt{2}$	$\sqrt{50}$	$\dfrac{1}{8}$

Playing the Game

Put the cards together to form a deck and shuffle the deck. Then pass out the cards so that each player has the same number of cards. You may end up with a few cards left over. Put these aside. Now you can begin playing the game.

To play a hand, have each player lay down a card. Then have the group order the numbers on the cards from greatest to least. Record the ordering of the numbers on a sheet of paper. The person who has the greatest number earns 6 points, the person who has the next greatest number earns 5 points, the person who has the third greatest number earns 4 points, and so on. After recording the number of points earned by each player, put the used cards aside. Continue to play hands and record the orderings of the numbers and the scores. The game is over when the players are out of cards. The winner is the player with the greatest number of points.

Teacher's Notes for Card Game Project

For use after Chapter 9

Project Goals
- Find and approximate square roots.
- Compare and order real numbers.

Managing the Project

Classroom Management This project will work well for groups of 4 to 6 people. You can make the game last longer by having the students reshuffle the used deck and redistribute the cards. You can also extend the game by having the students create their own cards to add to the deck.

The difficulty of the game can be changed by either allowing the use of calculators or not. If you do not allow the use of calculators, suggest that the students estimate the radicals to the nearest tenth.

Rubric for Project The following rubric can be used to assess student work.

4 The students keep a neat record of their work. All real numbers are correctly ordered. The score of the game is neatly and accurately kept.

3 The students keep a neat record of their work. Most of the real numbers are correctly ordered. The score of the game is neatly and accurately kept.

2 The students' record of their work is sloppy. The students make some errors when ordering the real numbers. The scoring of the game may not be accurate.

1 The students' record of their work is sloppy. The students make many errors when ordering the real numbers. The scoring of the game is not accurate.

CHAPTER 9

Independent Extra Credit Project: Cliff Diving

For use after Chapter 9

Objective Find the height of a cliff and the heights of a cliff diver.

Materials pencil, paper

Investigation *Getting Going* You work for a sports magazine and are writing an article on cliff diving. In your article, you want to include some photographs of cliff divers making their dives.

You are taking photos from a boat that is about 40 meters from the base of a cliff. To help you determine the height of the cliff, you use an instrument called a compass clinometer to measure the angle of elevation from your eye level to the top of the cliff. The distance from your eye level to the surface of the water is about 1.8 meters.

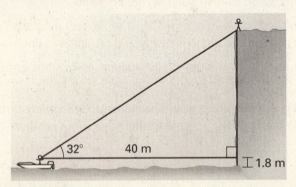

Questions

1. Find the height of the cliff. Round your answer to the nearest tenth of a meter.

2. Find the distance between you and the top of the cliff. Round your answer to the nearest tenth of a meter.

3. The entire path of the diver is about 2 meters from the face of the cliff. You take a photograph of the diver when the angle of elevation is 25° and when the angle of elevation is 20°. How far has the diver traveled during the time between the two photos? Round your answer to the nearest tenth of a meter.

4. If you look up at an object, the angle that your line of sight to the object makes with a horizontal line from your eyes is called the angle of elevation of the object. An angle of depression is formed in the same way but instead you are looking down at the object. A diver reaches the water and plunges to a certain depth. The angle of depression from your eye level to the diver is 9°. Find the depth of the diver. Round your answer to the nearest tenth of a meter.

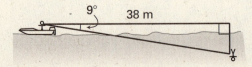

5. Find the height of the diver at several different angles of elevation.

Teacher's Notes for Cliff Diving Project

For use after Chapter 9

Project Goals
- Use the Pythagorean Theorem.
- Use the tangent to find lengths.

Managing the Project

Guiding Students' Work Remind students that when doing their calculations, to take into account that the path of the diver is 2 meters from the face of the cliff and your eye level is 1.8 meters above the surface of the water.

To help with their calculations, you can allow students to use calculators.

Alternative Approach You can explain to students how the angle of elevation in the drawing is calculated. Teach students how to find the angle of elevation using the TAN^{-1} keystroke on their calculators. For example, show them the distance from the cliff is 40 meters and the height of the cliff in regards to your eye level is 25 meters (from Question 1). So, $\tan^{-1}\left(\frac{25}{40}\right) \approx 32°$.

Rubric for Project The following rubric can be used to assess student work.

4 The student correctly finds the height of the cliff, the distance to the top of the cliff, the distance the diver traveled, and the depth of the diver. The student also calculates several heights of the diver at various angles of elevation. All of the calculations are correct with his/her work shown. The student's work is neat.

3 The student finds the height of the cliff, the distance to the top of the cliff, the distance the diver traveled, and the depth of the diver. There are some minor errors. The student also calculates several heights of the diver at various angles of elevation. Most of the calculations are correct with his/her work shown. The student's work is neat.

2 The student has some difficulty finding the height of the cliff, the distance to the top of the cliff, the distance the diver traveled, and the depth of the diver. There are some errors in the calculations. The student calculates several heights of the diver at various angles of elevation but there are some errors. Some of the calculations are incorrect and no work is shown. The student's work is a little sloppy.

1 The student has difficulty finding the height of the cliff, the distance to the top of the cliff, the distance the diver traveled, and the depth of the diver. There are several errors in the calculations. The student cannot calculate the heights of the diver at various angles of elevation. Some of the calculations are incomplete. The student's work is incomplete or sloppy.

Cumulative Practice

For use after Chapter 9

Evaluate the expression. (Lesson 1.3)

1. $15.3 - 4 \cdot 11$
2. $49 \div 7 + 16 \div 4$
3. $6(33 - 5^2)$
4. $\dfrac{3.9 + 8.2}{34 - 23}$

Evaluate the expression using the distributive property and mental math. (Lesson 2.2)

5. $5(99)$
6. $7(2.99)$
7. $-2(456)$
8. $8(4.1)$

Solve the equation. (Lessons 3.2, 3.3)

9. $60 = 4(2m + 1)$
10. $-2(-4x) = 3(6 + 2x)$
11. $-1(14 + 2c) = 5(c - 7)$

Use the LCD to determine which fraction is greater. (Lesson 4.4)

12. $\dfrac{2}{5}, \dfrac{3}{10}$
13. $\dfrac{7}{9}, \dfrac{6}{7}$
14. $\dfrac{10}{13}, \dfrac{11}{15}$
15. $\dfrac{19}{50}, \dfrac{25}{31}$

Solve the equation or inequality. (Lessons 5.6, 5.7)

16. $\dfrac{2}{3}x + \dfrac{2}{9} = \dfrac{5}{6}$
17. $\dfrac{7}{16}t - \dfrac{3}{16} = \dfrac{1}{16}$
18. $\dfrac{1}{5} + \dfrac{1}{8}d \leq \dfrac{19}{40}$
19. $x + \dfrac{1}{2} > \dfrac{2}{3}$

In Exercises 20–22, use the following information. Each letter in the word MATHEMATICAL is written on a separate slip of paper and placed in a hat. A letter is chosen at random from the hat. (Lesson 6.7)

20. What is the probability that the letter chosen is an M?
21. What is the probability that the letter chosen is a consonant?
22. What are the odds against choosing an A?

For an account that earns interest compounded annually, find the balance of the account. Round your answer to the nearest cent. (Lesson 7.7)

23. $P = \$3000$, $r = 5.5\%$, $t = 10$ years
24. $P = \$1280$, $r = 4.7\%$, $t = 3$ years
25. $P = \$740$, $r = 6.7\%$, $t = 4$ years

Cumulative Practice
For use after Chapter 9

Tell whether the ordered pair is a solution of the equation. (Lesson 8.2)

26. $y = 2x - 3$; $(0, -4)$

27. $2x + y = 1$; $(1, -1)$

28. $x + y = 5$; $(2, 3)$

29. $4x - y = 9$; $(2, 5)$

Graph the inequality in a coordinate plane. (Lesson 8.9)

30. $x - y \leq 3$

31. $y < 4x - 5$

32. $6x + 9y < -18$

33. $x \leq 3$

Find the unknown length. Write your answer in simplest form. (Lesson 9.3)

34.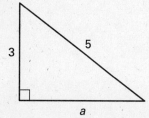

35.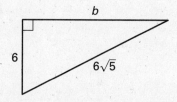

36.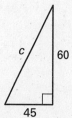

Find the value of x. Round to the nearest tenth. (Lessons 9.7, 9.8)

37.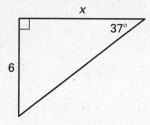

38.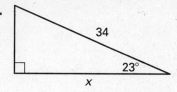

39.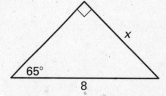

Answers

Lesson 9.1

Technology Activity
1. 7.21 **2.** 10.58 **3.** 22.36 **4.** 0.73
5. 5.53 **6.** 5.24 **7.** 5.87 **8.** 5.02 **9.** 0.70
10. 2.07 **11.** 4.06 **12.** 1.30

Practice A
1. ±7 **2.** ±8 **3.** ±10 **4.** ±14 **5.** ±20
6. ±50 **7.** 5 **8.** −10 **9.** 9 **10.** −11
11. −6 **12.** 3 **13.** 1.4 **14.** −2.8 **15.** 9.5
16. −7.3 **17.** 34.3 **18.** 3.8 **19.** 5.2 **20.** 7
21. −36 **22.** ±6 **23.** ±13 **24.** ±14
25. ±5.4 **26.** ±10.5 **27.** ±2.9 **28.** ±5
29. ±15.49 **30.** ±3.06 **31.** 5.5 feet
32. 22 feet **33.** yes

Practice B
1. ±6 **2.** ±19 **3.** ±27 **4.** ±33 **5.** ±70
6. ±100 **7.** 6 **8.** −9 **9.** 10 **10.** −12
11. 4 **12.** −2 **13.** 2.2 **14.** −3.5 **15.** 10.1
16. −8.6 **17.** 39.8 **18.** 5.3 **19.** 8 **20.** 9
21. −96 **22.** ±8 **23.** ±18 **24.** ±15
25. ±6.5 **26.** ±12.2 **27.** ±3.1 **28.** ±4
29. ±7.39 **30.** ±3.95 **31.** 172 feet
32. 9.3 miles

Practice C
1. 7 **2.** −10 **3.** 11 **4.** −12 **5.** −3 **6.** 3
7. 3.3 **8.** −5.4 **9.** 10.3 **10.** −14.7
11. 35.1 **12.** 3.2 **13.** 12 **14.** 13 **15.** −20
16. ±11 **17.** ±21 **18.** ±52 **19.** ±7.7
20. ±14.8 **21.** ±8.4 **22.** ±4.80
23. ±16.97 **24.** ±2.83 **25.** $x^2 = 0.0169$
26. a. 25.5 m **b.** 102 m **c.** no; 42 m

Study Guide
1. 5, −5 **2.** 3, −3 **3.** 10, −10 **4.** 11, −11
5. 4 **6.** 5 **7.** −8 **8.** 10 **9.** 3.9
10. −9.7 **11.** −1.9 **12.** 8.7 **13.** 20
14. −16 **15.** −12 **16.** ±5

Challenge Practice
1. −174 **2.** 560 **3.** −120 **4.** ±11.31
5. ±1.41 **6.** ±2.12 **7.** < **8.** = **9.** <
10. −4 or 10; First add 4 to each side to get $(3 - x)^2 = 49$. Then write 49 as 7^2. Because $(3 - x)^2 = 7^2$, $3 - x = 7$, or $x = -4$. Also write 49 as $(-7)^2$. Because $(3 - x)^2 = (-7)^2$, $3 - x = -7$, or $x = 10$.

Lesson 9.2

Practice A
1. $2\sqrt{5}$ **2.** $5\sqrt{6}$ **3.** $6\sqrt{7}$ **4.** $5\sqrt{7}$ **5.** $12\sqrt{3y}$
6. $4c\sqrt{3}$ **7.** $\frac{\sqrt{17}}{7}$ **8.** $\frac{\sqrt{23}}{9}$ **9.** $\frac{\sqrt{5}}{4}$ **10.** $\frac{2\sqrt{21}}{13}$
11. $\frac{\sqrt{p}}{10}$ **12.** $\frac{4x\sqrt{2}}{5}$ **13.** $10\sqrt{5}$ units
14. $2b\sqrt{15c}$ **15.** $20k\ell\sqrt{6}$ **16.** $2h\sqrt{30j}$
17. $\frac{10\sqrt{6x}}{3y}$ **18.** $\frac{m\sqrt{7}}{15n}$ **19.** $\frac{\sqrt{65y}}{4z}$ **20. a.** $2\sqrt{3\ell}$
b. about 45 mi/h **21.** $\frac{5\sqrt{3}}{4}$; about 2 seconds

Practice B
1. $3\sqrt{6}$ **2.** $4\sqrt{7}$ **3.** $4\sqrt{11}$ **4.** $6\sqrt{5}$
5. $3\sqrt{13f}$ **6.** $12y\sqrt{3}$ **7.** $\frac{2\sqrt{30}}{11}$ **8.** $\frac{\sqrt{3}}{3}$
9. $\frac{\sqrt{202}}{16}$ **10.** $\frac{8\sqrt{5}}{21}$ **11.** $\frac{v\sqrt{21}}{18}$ **12.** $\frac{\sqrt{94t}}{14}$
13. $10\sqrt{7}$ units **14.** $3d\sqrt{19c}$ **15.** $2m\sqrt{38n}$
16. $3xy\sqrt{14}$ **17.** $\frac{w\sqrt{23}}{7}$ **18.** $\frac{t\sqrt{5r}}{4}$ **19.** $\frac{2pq\sqrt{19}}{9}$
20. a. $3\sqrt{2\ell}$ **b.** about 50 mi/h
21. $\frac{7\sqrt{5}}{4}$; about 4 seconds

Practice C
1. $5\sqrt{14}$ **2.** $7\sqrt{15}$ **3.** $8\sqrt{7}$ **4.** $6\sqrt{3}$
5. $2xy\sqrt{46y}$ **6.** $3m\sqrt{3mn}$ **7.** $\frac{\sqrt{62}}{19}$ **8.** $\frac{\sqrt{17}}{7}$
9. $\sqrt{3}$ **10.** $\frac{3\sqrt{15}}{16}$ **11.** $\frac{4\sqrt{6t}}{17r}$ **12.** $\frac{2b\sqrt{30b}}{13a}$
13. $100\sqrt{5}$ units **14.** $3y\sqrt{21xz}$ **15.** $2mnp\sqrt{29}$
16. $2ac\sqrt{39ab}$ **17.** $\frac{\sqrt{11f}}{4d}$ **18.** $\frac{\sqrt{2}}{3x}$ **19.** $\frac{2w\sqrt{11}}{11}$
20. $\frac{\sqrt{345}}{2}$; about 9 seconds
21. $2.46\sqrt{6}$; about 6 miles **22.** $2\sqrt{7}, 4\sqrt{2}, 6$

Lesson 9.2 continued

Study Guide

1. $2\sqrt{6}$ 2. $10\sqrt{3}$ 3. $5\sqrt{11}$ 4. $-10\sqrt{10}$
5. $2y\sqrt{23}$ 6. $2\sqrt{3x}$ 7. $-3\sqrt{2t}$ 8. $-3k\sqrt{3}$
9. $-\dfrac{2\sqrt{10}}{9x}$ 10. $\dfrac{3\sqrt{3}}{50t}$ 11. $-\dfrac{3\sqrt{h}}{4}$
12. $-\dfrac{5a\sqrt{2}}{17}$ 13. about 3 seconds

Challenge Practice

1. $\dfrac{2}{9x}\sqrt{7}$ 2. $\dfrac{5}{8b}\sqrt{3a}$ 3. $\dfrac{8m}{3p}\sqrt{6n}$ 4. $2x\sqrt{2x}$
5. $2y^2\sqrt{3y}$ 6. $\dfrac{2m}{3n^2}\sqrt{m}$ 7. $4xy\sqrt{13}$ 8. $3\sqrt{2}$
9. 15 10. $2\sqrt{6}, 3\sqrt{3}, \sqrt{30}$

Lesson 9.3

Practice A

1. 10 2. 17 3. $\sqrt{29}$ 4. $16\sqrt{2}$ 5. $5\sqrt{5}$
6. $6\sqrt{15}$ 7. no 8. yes 9. no 10. yes
11. no 12. yes 13. 21 14. 40 15. 12
16. 27 17. 53 18. 65 19. a. Check figures.
b. $4^2 + x^2 = 12^2$ c. $x = \sqrt{128} = 8\sqrt{2}$ d. 11 ft
20. 13.9 in.

Practice B

1. 26 2. 25 3. $2\sqrt{10}$ 4. $3\sqrt{91}$ 5. $6\sqrt{5}$
6. $\sqrt{101}$ 7. yes 8. no 9. yes 10. no
11. no 12. no 13. 40 14. 42 15. 33
16. 24 17. 85 18. 65 19. 6.6 yd
20. 60.2 ft 21. 10.09 ft 22. 9.60 cm
23. 3.42 m

Practice C

1. 34 2. $6\sqrt{17}$ 3. $2\sqrt{281}$ 4. $15\sqrt{5}$
5. $3\sqrt{39}$ 6. $2\sqrt{70}$ 7. yes 8. no 9. yes
10. no 11. no 12. no 13. 48 14. 30
15. 12 16. 20 17. 68 18. 100
19. 127.3 feet 20. yes 21. 6.36 m

Study Guide

1. $4\sqrt{5}$ 2. 12 3. $5\sqrt{3}$
4. yes 5. no 6. no

Challenge Practice

1. 0.5 2. 0.5 3. $\sqrt{1.25}$ 4. 19 cm 5. 2.92
6. 5 7. 11 8. 6, 8, 10 9. 11 in.^{2}

Lesson 9.4

Practice A

1. no 2. rational; irrational 3. rational
4. rational 5. irrational 6. rational
7. irrational 8. rational 9. rational
10. rational 11. = 12. < 13. >
14. $\dfrac{\sqrt{16}}{2}, \sqrt{\dfrac{9}{2}}, \sqrt{10}, 3.4$ 15. $\dfrac{\sqrt{24}}{2}, \dfrac{19}{6}, \sqrt{\dfrac{36}{3}}, \sqrt{14}$
16. $3\sqrt{8}, \sqrt{75}, 9, 7\sqrt{2}$
17. $-2\sqrt{37}, -11, -\dfrac{21}{2}, -\sqrt{110}$
18. Sample answer: $\sqrt{83}$
19. rational; $s = \sqrt{A} = \sqrt{\dfrac{9}{16}} = \dfrac{3}{4}$, which is a
rational number. 20. 55.6 in.

Practice B

1. rational 2. irrational 3. rational
4. irrational 5. rational 6. irrational
7. irrational 8. rational 9. > 10. =
11. > 12. $\sqrt{1.1}, \dfrac{\sqrt{10}}{3}, \dfrac{12}{11}, \sqrt{\dfrac{10}{3}}$
13. $\sqrt{\dfrac{81}{5}}, \sqrt{17}, 4.2, \dfrac{17}{4}$
14. $\dfrac{2}{3}, \sqrt{\dfrac{5}{9}}, 0.7514, 0.\overline{75}, \sqrt{\dfrac{7}{9}}$
15. $\dfrac{\sqrt{75}}{2}, \sqrt{\dfrac{55}{2}}, 5.5, \sqrt{31}, \dfrac{28}{5}$
16. rational 17. irrational; 1.77 seconds
18. Yes; the storage compartment is 1.5 feet wide and 1.5 feet deep, so an 18 inch pizza should just fit. 19.

Practice C

1. rational 2. rational 3. rational
4. irrational 5. rational 6. rational
7. irrational 8. rational 9. > 10. =
11. < 12. $\sqrt{44}, 6.\overline{63}, \dfrac{41}{6}, 4\sqrt{3}$
13. $\sqrt{27}, \dfrac{21}{4}, 5.318, 3\sqrt{\dfrac{22}{7}}$

Lesson 9.4 continued

14. $-\frac{81}{11}, -\sqrt{54}, -2\sqrt{11}, -\frac{4\sqrt{37}}{5}$

15. $\frac{\sqrt{15}}{2}, 2.\overline{47}, \frac{57}{23}, \sqrt{\frac{15}{2}}, 3\sqrt{0.9}$

16. yes 17. The rectangle; its perimeter is 66 feet and the perimeter of the square is about 58.9 feet. 18. Yes; it takes about 3.67 seconds for the rock to hit the surf.

19. No. Sample answer: $\frac{5\sqrt{5}}{2}$ is an irrational number. An irrational number is a nonterminating decimal, so you cannot write its exact value as a decimal. 20. Sample answer: Draw a right triangle with both legs of length $\frac{2}{3}$ units. Start one leg at the origin and draw a straight line that is $\frac{2}{3}$ unit long. Draw a vertical line $\frac{2}{3}$ unit from the end of the first leg. Connect the end of this leg to the origin to form the hypotenuse. Then use a compass to transfer the length of the hypotenuse of the triangle to a number line.

Study Guide

1. rational 2. rational 3. rational
4. irrational 5. < 6. < 7. > 8. >
9. $\sqrt{21}, \frac{32}{5}, 5\sqrt{2}, 7.1$
10. $-2\sqrt{2}, -\sqrt{7}, -\frac{13}{5}, -2.3$
11. 23 miles per hour 12. 73 miles per hour
13. 45 miles per hour 14. 65 miles per hour

Challenge Practice

1. $\sqrt{17}, 4.15, 3\sqrt{2}, \frac{48}{11}, 4\frac{3}{5}, 4.7$
2. $-3\frac{4}{5}, -\frac{187}{50}, -3.5, -3.14, -2\sqrt{2}, -\sqrt{7}$
3. $4\sqrt{5}, 8.98, \sqrt{83}, 9\frac{1}{8}, \frac{183}{20}, 9.27$
4. $-0.87, -\frac{3}{4}, -0.71, -\sqrt{\frac{1}{2}}, -\sqrt{0.25}, -\frac{7}{16}$
5. Sample answer: $\frac{\sqrt{5}}{6}$
6. Sample answer: $\sqrt{31}$
7. Sample answer: $-\sqrt{18}$
8. Sample answer: sometimes; $\frac{\sqrt{2}}{\sqrt{2}} = 1$ is rational, but $\frac{\sqrt{15}}{\sqrt{3}} = \sqrt{5}$ is irrational.

Lesson 9.5

Activity Master

1. $4\sqrt{2}$ 2. $\sqrt{10}$ 3. 13 4. $3\sqrt{5}$
5. $C(x_1, y_2)$ or $C(x_2, y_1)$
6. Sample Answer:
$AB = \sqrt{(y_1 - y_2)^2 + (x_1 - x_2)^2}$

Practice A

1. $\sqrt{26}$ 2. 3 3. $\sqrt{17}$ 4. $\sqrt{26}$ 5. $2\sqrt{13}$
6. $\sqrt{89}$ 7. $3\sqrt{5}$ 8. $\sqrt{37}$ 9. $\sqrt{265}$
10. $2\sqrt{41}$ 11. $(3, -3)$ 12. $\left(-\frac{3}{2}, -6\right)$
13. $-\frac{3}{7}$ 14. undefined 15. $\left(-3, -\frac{1}{2}\right)$
16. $(3, -1)$ 17. $(5, 5)$ 18. $\left(-\frac{1}{2}, -1\right)$
19. $\left(-7\frac{1}{2}, -3\frac{1}{2}\right)$ 20. $\left(2, -\frac{1}{2}\right)$ 21. -1
22. $-\frac{3}{8}$ 23. $\frac{2}{3}$ 24. -6 25. $-\frac{8}{3}$ 26. -4
27. a. $\sqrt{10}$ units   b. $\sqrt{10}$ units

Practice B

1. $5\sqrt{2}$ 2. $\sqrt{26}$ 3. $\sqrt{29}$ 4. $\sqrt{73}$ 5. $\sqrt{586}$
6. $2\sqrt{17}$ 7. $\sqrt{94.25}$ 8. $\sqrt{31.25}$ 9. 5
10. $6\sqrt{5}$ 11. $\left(0, 3\frac{1}{2}\right)$ 12. $\left(2, -\frac{1}{2}\right)$
13. $(-2, -2)$ 14. $(-4, 2)$ 15. 3 16. $-\frac{5}{3}$
17. $\frac{7}{3}$ 18. $-\frac{1}{5}$ 19. $(-4, -3)$ 20. $(1, 1)$
21. $(3.5, -1.5)$ 22. $(2.3, 2.8)$ 23. $(1, 4.5)$
24. $\left(14, -2\frac{1}{4}\right)$ 25. $\frac{9}{2}$ 26. -1 27. 10
28. -0.16 29. $-\frac{22}{7}$ 30. $\frac{3}{115}$
31. a.   b. $M\left(2, 1\frac{1}{2}\right)$

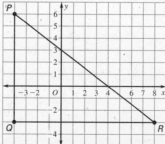

Lesson 9.5 continued

Practice C

1. $2\sqrt{17}$ 2. $\sqrt{61}$ 3. $2\sqrt{2}$ 4. $\sqrt{61}$ 5. $\sqrt{41.6}$
6. $\sqrt{15.88}$ 7. $\sqrt{39.88}$ 8. $\sqrt{100.84}$
9. $\sqrt{328.96}$ 10. $\sqrt{164.88}$ 11. $\left(-2, 5\frac{1}{2}\right)$
12. $\left(1\frac{1}{2}, -5\frac{1}{2}\right)$ 13. $\left(-5, -4\frac{1}{2}\right)$ 14. $\left(7, 3\frac{1}{2}\right)$
15. $\frac{7}{2}$ 16. $-\frac{1}{5}$ 17. $(-0.6, 1.8)$
18. $(-8.3, 12.7)$ 19. $(2.3, -5.4)$
20. $(7.5, 13.2)$ 21. $(17.7, 17.8)$
22. $(-13.9, 14.5)$ 23. $-\frac{95}{73}$ 24. $\frac{111}{46}$ 25. $-\frac{9}{8}$
26. $5\sqrt{2}$ feet 27. $\overline{AB}$ 28. $B(2, -4)$
29. $B(1, 4)$
30. 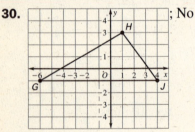 ; No

Study Guide

1. $7\sqrt{2}$ 2. $\sqrt{122}$ 3. $2\sqrt{85}$ 4. $\sqrt{10}$ miles
5. $\sqrt{13}$ miles 6. $2\sqrt{10}$ miles 7. $\left(6, 4\frac{1}{2}\right)$
8. $\left(1, 4\frac{1}{2}\right)$ 9. $\left(-6, -7\frac{1}{2}\right)$ 10. -1
11. $\frac{1}{3}$ 12. $-\frac{5}{6}$

Real-World Problem Solving

1. 100 ft 2. (60, 60) 3. 50 ft; 98 ft 4. 40 ft

Challenge Practice

1. not a right triangle 2. not a right triangle

3.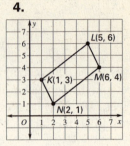

area: 100 units²;
perimeter: 40 units

4.

area: $5\sqrt{5}$ units²;
perimeter: $10 + 2\sqrt{5}$ units

5. 5 6. -6 7. 2 8. segment 1 endpoints: (0, 0) and (3, 2); segment 2 endpoints: (3, 2) and (6, 4); segment 3 endpoints: (6, 4) and (9, 6); segment 4 endpoints: (9, 6) and (12, 8) 9. ± 2

Lesson 9.6

Practice A

1. $3\sqrt{2}$ 2. $8\sqrt{2}$ 3. 10 4. $7\sqrt{3}$ 5. $9\sqrt{3}$
6. 12 7. The length of the hypotenuse is the product of the length of a leg and $\sqrt{2}$.
8. 10 in. 9. $5\sqrt{3}$ in. 10. $x = 12, y = 12\sqrt{2}$
11. $x = 18\sqrt{2}, y = 18$ 12. $x = 4\sqrt{2}, y = 4$
13. $x = 5\sqrt{3}, y = 10$ 14. $x = 8\sqrt{3}, y = 8$
15. $x = 22, y = 11\sqrt{3}$ 16. 11 ft

Practice B

1. $9\sqrt{2}$ 2. $15\sqrt{2}$ 3. 17 4. $10\sqrt{3}$ 5. $3.5\sqrt{3}$
6. 5 7. The length of the hypotenuse is twice the length of the shorter leg.
8. $x = 13\sqrt{2}, y = 13$ 9. $x = 25\sqrt{2}, y = 25$
10. $x = 7.3, y = 7.3\sqrt{2}$ 11. $x = 15\sqrt{3}, y = 30$
12. $x = 17, y = 8.5\sqrt{3}$
13. $x = 10.5\sqrt{3}, y = 10.5$ 14. 5 ft; 7 ft

Practice C

1. 16 2. $8.5\sqrt{2}$ 3. $2.25\sqrt{2}$ 4. $6.5\sqrt{3}$
5. $12.5\sqrt{3}$ 6. 7.2 7. $x = 34\sqrt{2}, y = 34$
8. $x = 6.5\sqrt{2}, y = 6.5$ 9. $x = 8.9, y = 8.9$
10. $x = 27, y = 13.5\sqrt{3}$
11. $x = 14.5\sqrt{3}, y = 14.5$
12. $x = 6.4, y = 12.8$

Lesson 9.6 continued

13. *Sample answer:* Fold the triangle in half to create two congruent 30°-60°-90° triangles. The height of the sign can then be found by finding the length of the longer leg of either triangle, which is $18\sqrt{3} \approx 31$ inches.

14. $AC = 11\sqrt{3}$ m, $BC = 22$ m, $XZ = 6\sqrt{3}$ m, $YZ = 12$ m

Study Guide

1. $x = 7, y = 14$ **2.** $x = 6, y = 6\sqrt{3}$
3. $x = 6, y = 3\sqrt{2}$ **4.** $x = 8$ ft, $y = 8\sqrt{3}$ ft

Challenge Practice

1. $x = 5\sqrt{2}$ **2.** $x = \frac{4\sqrt{3}}{3}, y = \frac{8\sqrt{3}}{3}$
3. $x = \frac{13\sqrt{3}}{2}, y = \frac{13}{2}$ **4.** $x = 8\sqrt{3}, y = 16\sqrt{3}$
5. $x = 5\sqrt{3}$ **6.** $x = \frac{44\sqrt{3}}{9}, y = \frac{22\sqrt{3}}{9}$
7. $8\sqrt{3}$ **8.** $\frac{3\sqrt{6}}{2}$ **9.** $4\sqrt{6}$ **10.** $4\sqrt{2}, 4\sqrt{2}, 8$;
Graph the triangle, then find the length of one of the legs using the distance formula. Then find the other lengths by using the angles of the triangle.

Lesson 9.7

Practice A

1. tangent   **2.** trigonometric   **3.** $\frac{28}{45}$ **4.** $\frac{12}{35}$
5. $\frac{99}{20}$ **6.** 0.3839   **7.** 28.6363   **8.** 1.8807
9. 0.0349   **10.** 1.1918   **11.** 3.0777
12. 0.8391   **13.** 0.4663   **14.** 0.7002   **15.** 46.9
16. 12.7   **17.** 35.8   **18.** 119 ft   **19.** The side opposite the angle is shorter than the side adjacent to the angle.

Practice B

1. $\tan B = \frac{8}{15}$; $\tan A = \frac{15}{8}$ **2.** $\tan D = \frac{5}{12}$;
$\tan E = \frac{12}{5}$ **3.** $\tan P = \frac{4}{3}$; $\tan Q = \frac{3}{4}$ **4.** 0.6249
5. 2.4751   **6.** 0.9325   **7.** 4.0108   **8.** 0.2493
9. 7.1154   **10.** 0.4040   **11.** 1.4826
12. 0.8098   **13.** 21.0   **14.** 64.3   **15.** 39.7
16. 1385.6 ft   **17.** about 220 ft

Practice C

1. $\tan A = \frac{3}{4}$; $\tan B = \frac{4}{3}$
2. $\tan D = \frac{28}{45}$; $\tan F = \frac{28}{16.5}$;
$\tan \angle GEF = \frac{16.5}{28}$; $\tan \angle DEG = \frac{45}{28}$
3. $\tan R = \frac{15}{8}$; $\tan S = \frac{8}{15}$; $\tan \angle STU = \frac{15}{8}$
4. 1.4550   **5.** 0.6899   **6.** 0.3859   **7.** 0.0052
8. 0.2107   **9.** 572.9572   **10.** 0.9004
11. 0.3249   **12.** 0.7536   **13.** 92.2   **14.** 39.0
15. 50.0   **16.** $x \approx 104; y \approx 111$ **17.** 22.4 ft
18. No limit; *Sample answer:* As an angle gets closer and closer to 90°, its tangent gets larger and larger.

Study Guide

1. $\frac{4}{3}$ **2.** 0.1763   **3.** 0   **4.** 1.5399   **5.** 0.7002
6. 14.6   **7.** 48.5

Real-World Problem Solving

1–2.

Building	Height (ft)	Height in Scale Model (in.)
Library	83	14
Elementary School	25	4
Middle School	58	10
High School	67	11
Town Hall	35	6
Bank	30	5
Supermarket	28	5

Challenge Practice

1. $\sqrt{19}$; $\tan A = \frac{\sqrt{19}}{9}$ **2.** $\sqrt{69}$; $\tan A = \frac{10\sqrt{69}}{69}$
3. $\sqrt{39}$; $\tan A = \frac{5\sqrt{39}}{39}$ **4.** 167.8 units^2
5. 278.4 units^2 **6.** 414.6 units^2 **7.** 9.1 in.^{2}
8. 3.3 mm^2 **9.** 41.9 cm^2

Lesson 9.8

Practice A

1. $\sin A = \frac{12}{13}$; $\cos A = \frac{5}{13}$; $\sin B = \frac{5}{13}$; $\cos B = \frac{12}{13}$ 2. $\sin S = \frac{4}{5}$; $\cos S = \frac{3}{5}$; $\sin T = \frac{3}{5}$; $\cos T = \frac{4}{5}$ 3. $\sin M = \frac{7}{25}$; $\cos M = \frac{24}{25}$; $\sin N = \frac{24}{25}$; $\cos N = \frac{7}{25}$ 4. 0.5 5. 0.3256
6. 0.9135 7. 0.2924 8. 0.7986 9. 0.9962
10. 0.9744 11. 0.9205 12. 0.9986 13. 8.5
14. 13.4 15. 21.6 16. 94.4 17. 35.0
18. 18.4 19. 59 m

Practice B

1. $\sin A = \frac{20}{29}$; $\cos A = \frac{21}{29}$; $\sin B = \frac{21}{29}$; $\cos B = \frac{20}{29}$ 2. $\sin S = \frac{9}{41}$; $\cos S = \frac{40}{41}$; $\sin T = \frac{40}{41}$; $\cos T = \frac{9}{41}$ 3. $\sin M = \frac{4}{5}$; $\cos M = \frac{3}{5}$; $\sin N = \frac{3}{5}$; $\cos N = \frac{4}{5}$ 4. 0.7071
5. 0.9563 6. 0.4226 7. 0.7986 8. 0.5736
9. 0.5299 10. 0.1392 11. 0.4695
12. 0.6947 13. 19.3 14. 7.3 15. 66.5
16. 57.6 17. 66.4 18. 46.8 19. 22.9 m

Practice C

1. $\sin M = \frac{5}{13}$; $\cos M = \frac{12}{13}$; $\sin N = \frac{12}{13}$; $\cos N = \frac{5}{13}$ 2. $\sin S = \frac{7}{25}$; $\cos S = \frac{24}{25}$; $\sin T = \frac{24}{25}$; $\cos T = \frac{7}{25}$ 3. $\sin A = \frac{20}{29}$; $\cos A = \frac{21}{29}$; $\sin B = \frac{21}{29}$; $\cos B = \frac{20}{29}$ 4. 0.8192
5. 0.1045 6. 0.6157 7. 0.1736 8. 0.8829
9. 0.6561 10. 0.3584 11. 0.8988
12. 0.6561 13. 19.1 14. 15.6 15. 4.2
16. 7.9 17. 118.8 18. 3.2 19. 0.8 m, 2.2 m
20. No; *Sample answer:* The hypotenuse is always greater than the length of either leg, so the ratios will always be less than 1.

Study Guide

1. $\sin N = \frac{24}{25}$; $\cos N = \frac{7}{25}$ 2. 0.2924
3. 0.9205 4. 0.8192 5. 0.0872
6. 20.4 7. 7.1

Challenge Practice

1. $4\sqrt{2}$; $\sin A = \frac{7}{9}$; $\cos A = \frac{4\sqrt{2}}{9}$
2. $\sqrt{161}$; $\sin A = \frac{\sqrt{161}}{15}$; $\cos A = \frac{8}{15}$
3. $3\sqrt{31}$; $\sin A = \frac{11}{20}$; $\cos A = \frac{3\sqrt{31}}{20}$
4. *Sample answer:* 3, 4, 5
5. *Sample answer:* 8, 13, $\sqrt{105}$
6. *Sample answer:* 7, 15, $4\sqrt{11}$
7. *Sample answer:* $\cos A$; The side opposite angle A will be the shortest length of the triangle. Because $\cos A = \frac{\text{adjacent}}{\text{hypotenuse}}$ and $\sin A = \frac{\text{opposite}}{\text{hypotenuse}}$ and the length of the leg opposite angle A is less than the length of the leg adjacent to angle A, $\cos A > \sin A$.
8. *Sample answer:* $\sin A$; The side adjacent to angle A will be the shortest length of the triangle. Because $\cos A = \frac{\text{adjacent}}{\text{hypotenuse}}$ and $\sin A = \frac{\text{opposite}}{\text{hypotenuse}}$ and the length of the leg adjacent to angle A is less than the length of the leg opposite angle A, $\sin A > \cos A$.

Review and Projects

Chapter Review Games and Activities

1. TRIGONOMETRIC RATIO
2. RADICAL EXPRESSION
3. LEG
4. REAL NUMBER
5. COSINE
6. SQUARE ROOT
7. PYTHAGOREAN THEOREM
8. MIDPOINT
9. TANGENT
10. PERFECT SQUARE
11. SINE
12. HYPOTENUSE
13. IRRATIONAL NUMBER

1–13. Examples may vary.

Review and Projects continued

Real-Life Project

1.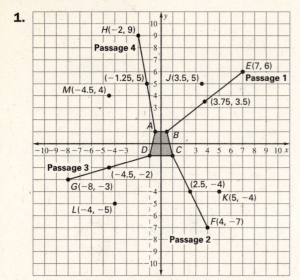

2. 7.1 units 3. See coordinate plane above for each passage location. **Passage 1:** 8.2 units; **Passage 2:** 6.7 units; **Passage 3:** 7.3 units; **Passage 4:** 8.1 units; From shortest to longest, the order is Passage 2, Passage 3, Passage 4, and Passage 1.

4. See coordinate plane above for each midpoint. **Passage 1:** (3.75, 3.5); **Passage 2:** (2.5, −4); **Passage 3:** (−4.5, −2); **Passage 4:** (−1.25, 5)

5. See coordinate plane above. 6. Each artifact was found 5 units from the closest corner of the main chamber.

Cooperative Project

Check students' work.

Independent Project

1. 26.8 m 2. 47.2 m 3. 3.9 m 4. 6.0 m
5. Check student's work.

Cumulative Practice

1. −28.7 2. 11 3. 48 4. 1.1 5. 495
6. 20.93 7. −912 8. 32.8 9. 7 10. 9
11. 3 12. $\frac{2}{5} > \frac{3}{10}$ 13. $\frac{7}{9} < \frac{6}{7}$ 14. $\frac{10}{13} > \frac{11}{15}$
15. $\frac{19}{50} < \frac{25}{31}$ 16. $\frac{11}{12}$ 17. $\frac{4}{7}$ 18. $d \leq 2\frac{1}{5}$

19. $x > \frac{1}{6}$ 20. $\frac{1}{6}$ 21. $\frac{7}{12}$ 22. $\frac{4}{1}$
23. $5124.43 24. $1469.10 25. $959.16
26. not a solution 27. solution 28. solution
29. not a solution

30. 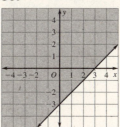   31.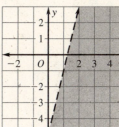

32.    33.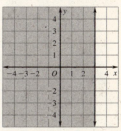

34. 4 35. 12 36. 75 37. 8 38. 31.3
39. 7.3